FAX

An Impartial

Guide to

Buying & Using

Facsimile

Machines

Daniel Fishman

Elliot King

THE BOOK OF FAX — Second Edition
An Impartial Guide to Buying and Using Facsimile Machines

Copyright © 1990 Daniel Fishman and Elliot King

Library of Congress Cataloging-in-Publication Data

Fishman, Daniel.
    The book of FAX.

    1. Facsimile transmission—Equipment and supplies—
Purchasing.   I. King, Elliot.   II. Title.
TK6710.F57 1990   621.382'35   89-70641
ISBN 0-940087-42-1

Cover design by Holly Russell, Durham, NC

Cover photography by Henderson-Muir Photography, Raleigh, NC

Illustrations by Bob Murray, Durham, NC

Desktop publishing by Pixel Plus, Chapel Hill, NC

Book design by Karen Wysocki, Ventana Press

Editorial Staff: Jeffrey Qualls, Terry Patrickis, Elizabeth Shoemaker,
                Marion Laird

Second Edition, First Printing
Printed in the United States of America

Ventana Press, Inc.
P.O. Box 2468
Chapel Hill, NC 27515
919/942-0220
919/942-1140 Fax

Limits of Liability and Disclaimer

ABOUT THE AUTHORS

Daniel Fishman is founder of Micro/Research, a computer and technology consulting firm in Monrovia, California, that specializes in integrating new technology to small and medium-sized businesses.

Daniel Fishman
Micro/Research
730 East Cypress Ave.
Monrovia, CA 91016
818/359-3503

Elliot King teaches communications at the University of California, San Diego, and has published more than 150 articles in computing magazines, including several on fax technology.

Elliot King
199 N. Camino Real
Suite F 272
Encinitas, CA 92024
619/436-4174

ACKNOWLEDGMENTS

The authors wish to thank Jan Ozer of Asher Technologies for his assistance, Jeffrey Charlston of Charlston, Revich & Williams for his research into the legal aspects of fax, Mark Chodos for his many hours of research, and Bill Gladstone for his help in initiating the project. We also wish to express our thanks to Anita and Sharon for their help and support during the writing of this book.

TRADEMARKS

CONTENTS

INTRODUCTION   **xxi**

THE INFORMATION MANAGEMENT
"BILL OF RIGHTS"   xxi

FAX: A NEW CHANNEL
OF COMMUNICATION   xxi

WHAT IS FAX?   xxii

WHO SHOULD READ THE BOOK OF FAX   xxiii

WHAT'S INSIDE   xxiii

HOW TO USE THE BOOK OF FAX   xxiv

DON'T BE LEFT BEHIND   xxiv

CHAPTER 1
WHY FAX?   **1**

WHY NOW?   1

MISCONCEPTIONS ABOUT FAX   2

COMPARING COMMUNICATIONS
OPTIONS   3

     The Mail   3

The Telephone 4

Overnight Delivery Service and Messengers 5

Telegram and Telex 6

Computer Networks 7

Fax 8

UNDERSTANDING YOUR
INFORMATION NEEDS 9

HOW COST-EFFECTIVE IS FAX? 10

Interpreting the Results 12

MOVING ON 13

CHAPTER 2
A LOOK AT FAX FEATURES 15

FINDING FEATURES TO SUIT YOUR NEEDS 15

BASIC FAX FEATURES 16

Compatibility 16

Resolution 19

Paper 22

Document Size 23

Unattended Operation 24

Transmission Time 24

Size and Weight 26

ADVANCED FEATURES 26

ADVANCED TRANSMISSION FEATURES 27

Built-in Dial 27

One-Touch Dialing 28

Speed Dialing 29

Automatic Redialing 30

Alternate-Number Dialing 31

Transmission Reservation 31

Delayed Transmission 32

Call Notification 32

Automatic Voice/Data Switch 32

Document Feeder 34

Automatic Paper-Cutting 34

Copier Option 35

Automatic Fallback 36

Transmit-Terminal Identification 36

Batch Index 37

Polling 38

Security 39

Delayed Transmission/Timer Transmission 40

Optical Mark Reader Sheets 41

Memory Features 41

Broadcast Transmission 42

Store and Forward 43

Substitute or Backup Reception 43

Memory Transmission 43

ADVANCED PRODUCTION FEATURES 44

Error Correction Mode 44

Image Enhancement Systems 44

Contrast Control 45

Oversized Document Reduction 45

Halftone Transmission   46

RS 232 Port or Interface   46

ADVANCED MANAGEMENT FEATURES   47

Transmission Report   47

Communications Journal   47

MOVING ON   49

CHAPTER 3
MATCHING FAX TO YOUR NEEDS   **51**

CHOOSING THE RIGHT FAX   51

PRICE VS. FUNCTION   53

ASSESSING THE FEATURES   54

Compatibility   54

Receiving Paper Capacity   55

Automatic Paper Cutter   56

Document Feeder Size   57

Telephone Handset   58

Document Sending Size   58

Copier Option   58

Resolution   59

Halftone Transmission   59

Superfast Mode   60

Delayed Transmission   61

Dialing Capabilities   61

Store and Forward   62

Broadcast Capability   62

Polling   63

Automatic Voice/Data Switch 63

Security and Secure Mailboxes 64

RS 232 Port 66

Activity Reports 66

MAJOR PRICE GROUPINGS 67

No-Frills Fax 67

Middle-Range Fax 68

Full-Featured Fax 70

Deluxe Fax 70

WHERE TO BUY 71

Manufacturers' Sales Representatives 71

Office Equipment Stores 72

Discount and Mail-Order Companies 72

FEATURES ASSESSMENT CHECKLIST 73

MOVING ON 79

CHAPTER 4
UP AND RUNNING **81**

A FAX MACHINE FRESH OUT OF THE BOX 81

Key Steps to a Smooth Transition 81

WHERE TO PUT IT? WHO WILL USE IT? 82

Location in Large Offices 82

Location in Smaller Companies 83

Rules of Thumb for Fax Machine Location 84

CHOOSING LONG-DISTANCE SERVICE 85

INSTALLING YOUR EQUIPMENT 87

The Manufacturer's Representative 88

Office Equipment Dealers   88

Mail Order or Discount Stores   89

AVOIDING POWER OUTAGE PROBLEMS   90

THE BUCK STOPS HERE   90

ADDITIONAL RESOURCES   90

THE FINAL STEPS   91

MOVING ON   91

CHAPTER 5
MANAGING YOUR FAX   **93**

INFORMATION MANAGEMENT   94

Entering the Fax Community   94

Getting the Word Out   95

Increasing Your Fax Audience   96

Checking Your Transmissions   97

Backing up Your Fax   100

The Right Tool for the Right Job   101

Protecting Confidential Information   102

Checking Your Incoming Messages   102

Routing and Storing Fax Messages   102

Protecting Your Fax Documents   103

Streamlining Fax Distribution   103

Fax Junk Mail   104

Internal Fax Misuse and Abuse   107

Legal Considerations   108

PERSONNEL MANAGEMENT   109

Do You Need an Operator?   109

Company Fax Directories 110

Encouraging Fax Use 110

Keeping the Machines Fit 111

Staying Up-to-Date 111

COST MANAGEMENT 112

SOME RULES FOR MANAGING FAX:
A SUMMARY 112

CHAPTER 6
PC/FAX: FRIEND OR FOE? **115**

COMBINING FAX WITH
PERSONAL COMPUTERS 115

ELECTRONIC MAIL VS. PC/FAX 116

UNDERSTANDING PC/FAX 117

Fax Boards 117

Optical Character Recognition (OCR)
Versus Scanners 118

Manipulating Fax Information 118

Fax Board Software 119

Fax Board Prices 120

COMPATIBILITY ISSUES 121

Other Hardware Considerations 122

RECEIVING MESSAGES VIA PC/FAX 122

PRINTING PC/FAX MESSAGES 124

INSTALLING PC/FAX 124

LEARNING TO USE PC/FAX 124

Differences in Fax Board Operations 125

Management Reporting Features   125
Other Software Features   125
Hard Copy Reproduction   126
Unit Sizes with PC/FAX Transmission   127

FAX BOARD NETWORKING   127

NEW DIRECTIONS   128
MCA/EISA   128
Binary File Transfer   128
Standard Communications Interface   129

THE BIG QUESTION: DO YOU
NEED PC/FAX?   129

PC/FAX ALTERNATIVES   131

CHAPTER 7
THE CELLULAR/FAX CONNECTION   **135**

ACOUSTICAL COUPLERS   136

CELLULAR JACKS   137

POWER SUPPLY   138

MOVING ON   138

CHAPTER 8
INNOVATIVE FAX USE   **141**

FAX ON THE ROAD   141

DISTRIBUTING INFORMATION   144

FAX FOR FUN   146

FAX MISCELLANY   147

MOVING ON   147

CHAPTER 9
A LOOK INTO THE FUTURE 149

RAPID BUT UNEVEN GROWTH 149

COMPATIBILITY 150

PRICING 150

ERROR CORRECTION 151

INCREASED PAPER CAPACITY, LASER
PRINTING AND BOND PAPER 151

INTEGRATION WITH OTHER OFFICE
EQUIPMENT 152

INCREASED MEMORY 153

NEW COMPUTER-BASED MACHINES 153

INTEGRATION WITH OTHER FORMS OF
COMMUNICATION 154

"FOURTH GENERATION" FAX 154

PORTABLE FAX 155

THE FUTURE OF FAX 155

APPENDIX A
TO LEASE OR TO BUY 157

APPENDIX B
COMMERCIAL FAX CENTERS 159

APPENDIX C
SELECTED FAX RESOURCES 161

Vendors/Manufacturers 161

Research 164

Accessories and Supplies   165

Publications   166

Directories   167

Associations   168

INDEX   **169**

THE INFORMATION MANAGEMENT "BILL OF RIGHTS"

In the early 19th century, Baron de Rothschild launched one of the great fortunes of Europe when he became the first Londoner to learn that Napoleon had been defeated at Waterloo. Armed with that important knowledge, he invested in the stock market and reaped great rewards.

Six months earlier, Andrew Jackson fought the Battle of New Orleans even though a peace treaty with the British had been signed. Because he hadn't received word from Washington about the treaty, the war dragged on needlessly.

Although the above examples are extreme, getting the *right* information at the *right* time to the *right* place in the *right* form offers a dramatic strategic advantage. On the other hand, being uninformed can be a recipe for trouble.

FAX: A NEW CHANNEL OF COMMUNICATION

We have long since outgrown the messenger service Baron de Rothschild used to such advantage. In many ways we've also outgrown telex, telephones and overnight mail delivery. Those who can effectively establish and manage a complete communications arsenal—telecommunications, computers and facsimile—enhance their chances to emerge as tomorrow's winners. To paraphrase Vince Lombardi, these days information isn't everything; it's the only thing.

WHAT IS FAX?

An innovative technology, contemporary facsimile equipment plays a key role in the information and communication revolution. In brief, fax (short for *facsimile*) equipment lets you duplicate almost any type of text or graphic information and transmit it over telephone lines to another fax machine.

To transmit information, fax machines scan documents and convert the information into an electronic format. That code is then sent via regular telephone lines to the receiving machine. There the information is decoded and printed, precisely reproducing the original document. At that point, the recipient has an exact duplicate of the original, while the sender keeps the original copy.

Unlike telephones, which usually transmit only sounds, fax transmits actual copy. Unlike computers, which transmit only files, fax transmits any graphic image including signatures, letterhead stationery and halftones. Unlike telex, which is slow and expensive, fax transmits vast amounts of information for about the price of a business phone call.

Fax works on a single international standard, which means that every machine can transmit to and receive from every other machine anywhere in the world. In fact, some countries rely more on fax than on telephones to send and receive information. For example, the Japanese currently have the largest installed base of fax users in the world. In Japan, to do business without a fax machine is like playing baseball without a mitt!

WHO SHOULD READ *THE BOOK OF FAX*

Anyone interested in fax technology, particularly those who want to buy a fax machine, will find *The Book of Fax* useful. It's written for nontechnical readers who need to learn the basics of fax equipment, and its use and management.

Because the authors and the publisher aren't affiliated with a manufacturer or other interest group, this book's impartial approach can help you objectively determine your needs—or whether you need fax technology at all.

For those who already own fax, *The Book of Fax* outlines ways to obtain the full benefit of fax technology in the most cost-effective way. Too often, when people buy new technical equipment, they use the easiest features and apply them only to existing methods of doing business. Because fax lets professionals communicate in ways that weren't possible in pre-fax times, this book explores many new fax management techniques, including some important "do's and don'ts."

WHAT'S INSIDE

The Book of Fax is a complete guide to purchasing, using and managing fax. In it you'll find

- ■ An introduction to facsimile technology—how it works and how it can help your business.

- A step-by-step analysis to determine whether you need fax.

- A comprehensive review of the major fax features—from the most common to the most advanced.

- A systematic method of identifying the right fax for your communications needs and your budget.

- A tested set of procedures for effectively and professionally managing fax, including some important "do's and don'ts."

- An overview of PC/FAX and its pros and cons.

- A resource guide to the fax community.

HOW TO USE *THE BOOK OF FAX*

If you don't yet own fax (or have just purchased one and aren't sure how it works), Chapters 1 through 3 will be of particular interest. You'll learn about fax features, how to assess them and determine which are best for your business and budget.

If you already own fax, you may want to skim the first three chapters for updated information, then read Chapters 4 through 8 for information on using and managing fax in your office or home. In each chapter, you'll find charts, worksheets or checklists that help you clarify needs and alternatives.

DON'T BE LEFT BEHIND

Fax is different from VCRs or telephone-answering machines. Professionals can't afford to be fax-illiterate. For many businesses, fax is as integral a part of communicating as the telephone.

As people discovered with telephone technology, it pays to stay informed. By being up-to-date you can make the best decisions and better communicate with other professionals, including consultants and sales personnel.

The Book of Fax is written to keep you at the forefront of this important new technology. After reading it, you'll be able to determine whether you need the strategic edge fax can offer. You also will be able to better manage your current communication channels. Let's get started.

Daniel Fishman
and Elliot King

CHAPTER 1

WHY FAX?

With all the talk about the paperless office of the future, why would anybody want to invest in a technology that generates more paper even faster? People are accustomed to picking up the telephone, using overnight delivery services, sending messengers and transmitting telexes. Until now, those channels have sufficed; so why invest in a new technology like fax?

Like it or not, paper (or *hardcopy*, as the jargon goes) is here to stay, at least for the foreseeable future. In fact, the widespread use of personal computers has actually *increased* the amount of paper generated by most businesses. The only difference is that now to remain competitive many businesses must send and receive information even faster.

With its unique ability to transmit existing graphics as well as text instantaneously, fax provides a communications capability that no other alternative offers.

WHY NOW?

Twenty years ago, fax was an expensive, slow, smelly, noisy technology, used only by international stockbrokers who hid the machines in their supply rooms and used them only when absolutely necessary. However, recent technological breakthroughs have made fax faster, quieter and less expensive, allowing businesses to take advantage of its unique abilities.

New, improved fax was snapped up quickly by businesses that relied heavily on international communications, had branch offices or needed to transmit visual information over long distances. But so many new uses have emerged that fax has quickly become a mainstream technology.

According to researchers, more than five million fax machines will be purchased by next year, and more than one million people will buy them in the next 12 months. Yet many people still know very little about the pros and cons of fax.

MISCONCEPTIONS ABOUT FAX

Because fax makes copies of original documents, some people think of it as little more than a remote photocopier.

Because fax can receive incoming transmissions unattended, some people think of it as little more than a telephone-answering device.

Because most fax machines are easy to operate for their most basic procedures, some people think of it as a simplified version of computer technology.

Finally, because fax provides instant transmission of documents, some people think of it as nothing more than a high-speed document delivery system.

Yes, fax makes duplicates of originals; yes, it can work unattended; yes, it transmits information electronically. But fax is more than a combination of existing technologies.

Fax is a strategic communications tool. Like all tools, it can be used and abused; it can work for or against you.

To get the most out of fax technology, you must understand both what it can and can't do. The challenge for fax users isn't just to transmit information quickly, but to ensure that the right information gets to the right person at the right time.

Whether you need fax equipment depends on how you use traditional methods of communication, the kind of information you deal with, its urgency and a host of other factors. By the end of this chapter, you should have a clear idea of your fax requirements, and how much time and money you might save by using fax. Chapters 2 and 3 describe basic and advanced fax features and help you match those features to your needs.

COMPARING COMMUNICATIONS OPTIONS

You'll better understand fax's advantages and disadvantages by comparing them with those of the traditional forms of communication.

Most ways information is distributed are so routine that you seldom think about them. You intuitively know whether to write a letter or make a phone call. Yet in determining whether to add fax to your communications system, you should take time to review the advantages and disadvantages of existing channels of communication.

The following comparative analysis should clarify the kinds of information you now send through traditional channels that may be better suited for fax transmission. Although this may seem an obvious exercise, it will help you visualize the role of fax in your overall communications functions.

The Mail

ADVANTAGES:

Though much maligned, the good old US Postal Service offers several advantages.

1. Mail can transmit information in many different forms and configurations, such as books, art and graphics, film and other media.

2. Mail provides complete coverage. It can be sent virtually anywhere in the world and requires no special equipment on the part of the recipient.

3. Mail is easy to use, and everybody understands how to send it. Most companies have efficient routines for processing and routing mail.

4. Mail is extremely private. Both custom and law dictate that only the addressee and those authorized by the addressee can open a letter or package.

5. Mail is inexpensive.

DISADVANTAGES:

1. Mail is slower than most alternatives.

2. Mail delivery may be erratic and out of the sender's control.

3. If your recipient gets a lot of mail every day, your message must compete for attention.

4. It is difficult to verify whether a message has been received.

5. Mail is time-consuming to produce and process.

The Telephone

In recent years, both technological advances and market competition have dramatically enhanced the usefulness of the telephone.

ADVANTAGES:

1. Telephone communication is instantaneous.

2. Telephone conversations are spontaneous and allow close interaction between participants.

3. Conversation is relatively private—both law and custom dictate that people not eavesdrop or make unauthorized recordings.

4. Like the mail, telephones provide very broad coverage—virtually everyone has a telephone.

5. Telephones are easy to use; little or no training is required.

DISADVANTAGES:

1. Telephone conversations require that both parties be available at the same time.

2. Except for short messages, information usually can't be transmitted through a third party for the recipient's later use.

3. You have no written verification of information communicated over the phone.

4. Because oral communication is a slow means of transmission, the telephone is often a poor medium for complex information (such as financial statements or engineering specifications).

5. Visual data, such as architectural plans, cannot be transmitted effectively over the telephone.

6. Employees can easily abuse the telephone by using it for personal calls. Unmonitored telephones can generate substantial bills.

Overnight Delivery Service and Messengers

As the need for high-speed communication has proliferated, services have arisen to move data physically from one place to another—within a city, across the country or around the world.

ADVANTAGES:

1. Overnight delivery enjoys many of the same advantages as mail.

2. It's fast.

3. Unlike the telephone, senders can calculate their costs in advance.

4. Overnight express service is a more dramatic, urgent form of communication than routine mail.

5. Local messenger services provide even better delivery schedules than do overnight services and convey an even greater sense of urgency.

DISADVANTAGES:

1. Overnight delivery (particularly for international delivery) and messenger services are expensive.

2. Although coverage is broad, overnight services don't reach all rural areas.

3. Messenger service usually is available only locally.

4. The sender is dependent upon pickup and delivery schedules controlled by a third party.

Telegram and Telex

Although domestic telegrams largely have been replaced by the telephone, telegrams still are used frequently for international business. Telex use, on the other hand, has increased rapidly both domestically and internationally.

ADVANTAGES:

1. Both telegrams and telex are dramatic forms of communication that command immediate attention.

2. They can be sent at any time and the recipient need not be present.

3. Both telex and telegrams have broad international coverage.

DISADVANTAGES:

1. Depending upon the volume of use, telex and telegrams can be very expensive.

2. Only small amounts of information can be sent in text form.

3. Telex and telegram messages are cumbersome to generate.

4. Telex and telegram messages are difficult to file.

5. Although you know that the message arrived at its destination, you can't be sure the person for whom it was intended actually received it.

6. Telex isn't private. Anyone near the telex terminal can read transmissions.

Computer Networks

ADVANTAGES:

1. Transmission is immediate.

2. Information is received in a form that can be revised easily.

3. Privacy and security systems can be built into the network.

4. The recipient need not be present.

5. The sender receives confirmation that the message has reached the intended destination.

6. Messages can be accessed quickly from remote locations.

DISADVANTAGES:

1. Coverage is limited to recipients who have appropriate equipment.

2. The technology is complex and costly.

3. Many computer networks cannot transmit graphics.

4. Standard communications protocols haven't been established, particularly for international communication.

5. Considerable (and often expensive) operator training is necessary.

Fax

Fax technology combines some of the best and worst features of the alternatives described above.

ADVANTAGES:

1. Like the telephone, fax transmits instantaneously.

2. Like mail, fax can send both text and graphics of virtually any length and complexity.

3. Like telex and telegrams, fax transmissions are dramatic, conveying a sense of urgency and demanding attention.

4. The recipient doesn't have to be present to receive a fax message.

5. The sender receives written confirmation that the message has been received.

6. Costs are moderate, amounting to little more than the telephone bill.

7. Messages arrive in a form that can be filed easily.

8. Fax technology is relatively simple, requiring very little training.

DISADVANTAGES:

1. Though fax use proliferates, some businesses still haven't adopted it.

2. Though basic fax is easy to use, advanced features

are sometimes complicated.

3. Fax is the only channel of communication for which the recipient and the sender incur supply and maintenance costs. Although most people buy fax to send information, they usually become recipients in short order.

4. Because sending costs are low, senders may transmit unsolicited fax messages demanding attention to matters the recipient might not wish to address.

5. Fax is so new that many users don't have routine procedures for handling incoming fax messages.

6. Fax messages aren't private; anyone at a fax terminal can read them.

7. The transmission quality of fax machines is uneven.

Some of the disadvantages (particularly the fact that some businesses don't yet have fax machines and the lack of internal procedures to handle fax traffic) will lessen as fax use grows. However, other disadvantages—such as costs to recipients and being at the mercy of unsolicited transmissions—will grow.

It's important to recognize that fax won't replace the channels of communication you now use. However, it can complement traditional communications methods, saving you both time and money and giving you an important competitive edge.

UNDERSTANDING YOUR INFORMATION NEEDS

You can better evaluate the kinds of information with which you work by answering six important questions about your communication needs:

1. With what kind of information do you work? Is it primarily text (e.g., accounting reports, spreadsheets, raw

data), or is it highly visual, containing many illustrations, photographs, charts or graphs?

2. How time-sensitive is the information you generate and receive? After you receive something, do you have three hours or 30 days to respond? Do you experience difficulty generating information and receiving responses within the required deadlines?

3. What volume of information do you send or receive?

4. Who produces the information with which you work? In general, is your information generated by one person working with a computer or calculator (e.g., accounting reports, spreadsheets, correspondence), or do you work with groups of people to produce company annual reports, advertising and public relations, or strategic plans? In other words, is most of the information you work with the product of an individual or a group?

5. Where is most of your information generated? Do you work primarily with internally generated information (e.g., profit and loss statements, internal memos, minutes to meetings), or do you work with information generated outside your location (e.g., correspondence, bids, contracts, purchase orders, invoices)?

6. Who reviews and receives the information you generate? Is your information circulated primarily to other people at your workplace or outside your location, such as remote offices and divisions, customers or suppliers? If your information generally is reviewed internally, where are the other people located who need to see what you produce?

HOW COST-EFFECTIVE IS FAX?

Your answers to the next series of questions can help you determine whether fax will be cost-effective for your business. Because you probably aren't familiar with all the fax features covered in Chapters 2 and 3, your cost and savings estimates will be crude. However, by completing the section below, you should gain some sense of your fax needs and potential savings.

How much money do you spend per month, excluding equipment and administrative charges, on the following:

$_____ Telephone usage

$_____ Postage

$_____ Telex and telegrams (include equipment and administrative charges in this figure)

$_____ Overnight delivery

$_____ Computer networks

(Equipment and administrative costs are included in the cost for telex and telegrams because fax could be a cost-effective replacement for telex. However, it won't replace your telephone system or your computer network; it probably won't even reduce the number of stations you need.)

Give the percentage of information you send via each channel that could be better handled by fax transmission.

%_____ Mail

%_____ Telephone

%_____ Telex and telegrams

%_____ Overnight delivery

%_____ Computer networks

Now calculate the total savings, based on the percentages above:

$_____ Telephone charges

$_____ Equipment costs

$_____ Administrative costs (outgoing)

$_____ Maintenance service and supplies

Would there be additional administrative costs in handling incoming fax messages? In other words, what volume of fax traffic do you anticipate? How much time would that take? How much would that time cost?

By calculating the percentage you would save on your traditional communication costs and subtracting your anticipated administrative costs, you can determine your approximate budget for fax equipment right now.

> FORMULA:
> Anticipated Savings - Anticipated Administrative Cost
> = Approximate Fax Equipment Budget

Interpreting the Results

If your anticipated fax equipment budget is $500 or more, you might want to continue to investigate fax. But even if your potential savings are small, you can get an accurate picture of how fax might fit into your business by imagining what you could do with fax capability that you can't do now.

Fax is an evolutionary technology. Once fax is in place, people often discover new ways to put it to use. Consequently, in determining whether you need fax, consider not only what you can do with it today but what you might do with it in the future—a key consideration in determining which fax features you need.

Item: A small company sells most of its products to a large corporation, which often is slow to pay its bills. When the small company calls to collect, the accounts payable supervisor denies they ever received the invoice. In response, the accounts receivable manager at the small company begins faxing a copy of the invoice immediately prior to calling. If the customer claims not to have a copy of invoice, the accounts receivable manager informs the customer that a copy of the invoice is in his or her fax machine at that very moment, thus cutting down the time it takes to collect payment.

Item: A company finds that whenever its salespeople exhibit at trade shows, they're virtually incommunicado. Moreover, it's often inconvenient for the salespeople to send orders written on the show floor back to headquarters promptly. The company includes a fax machine in its booth to send messages to its salespeople and receive orders immediately.

Item: Before and during the process of writing proposals, a consultant likes to receive extensive input from the client. He often spends hours on the telephone thrashing out the details. He shortens the process by developing written material on his own and faxing that information to the client before each phone call. Because the client has time to consider each version beforehand, telephone calls are shorter, and fewer are needed to develop the final proposal.

MOVING ON

By now, you should have some idea of how you would benefit from fax. The next two chapters provide an overview of how fax works and its basic and advanced features.

CHAPTER 2

A LOOK AT
FAX FEATURES

FINDING FEATURES TO SUIT YOUR NEEDS

Understanding the key characteristics and features available in different machines is the next step in exploring the potential of fax technology.

Like other advanced technology, fax equipment is available in a wide price range. Stripped-down units sell for less than $500. More sophisticated models cost more than $3,000. But price alone cannot be your major criterion. Low-end fax machines may not be suitable for the tasks you want to perform. On the other hand, the high-priced models may be needlessly complicated and packed with features you'll never use.

Not all fax equipment is equal, particularly in regard to your business. The features of any machine may be roughly categorized as either basic or advanced.

Basic features are shared by all fax machines, from standard workhorse models to the most complex, state-of-the-art, multipurpose equipment. In deciding which fax machine to buy or lease, you can only benefit from an understanding of the basic features.

Advanced features provide additional capabilities at additional cost. And though they sound wonderful in theory, many are still clumsy to use. Understanding the advanced features will help you determine the level of sophistication you need. Bear in mind that what might be a frivolous feature for one business can be vital for another.

Powerful advanced features also affect the price of a fax machine. Even if a low-end machine does have the features you want, those features may not be powerful enough to meet your needs. For example, a document feeder that can handle only five pages is of little use if you normally transmit longer documents.

So it's always wise to be realistic about what you want from your machine's features and spend the extra money for the capabilities and power you need.

BASIC FAX FEATURES

The first step in analyzing fax technology is to review the basic operational features all fax machines share that allow them to communicate with each other.

The most important feature is compatibility. Although fax technology enjoys internationally accepted communications standards, not all machines *use* the same standards.

Other key operational features to consider are the type of paper a machine uses, the size of documents it can send and receive, the quality or resolution of the image it can send and receive, its transmission speed, and the size and weight of the unit.

Compatibility

As with personal computers, compatibility (a machine's ability to send and receive material from another machine) is a crucial factor. Fortunately, international standards for fax technology have been established by the International Consultative Committee on Telegraphy and Telephony (CCITT), based in Geneva, Switzerland. As technology has improved, the CCITT has authorized improved communication standards as well.

Currently, four international standards exist for fax technology: Group 1, Group 2, Group 3 and Group 4 (which works with digital networks [ISDN]). The standards differ in the way information is scanned and coded, and the speed at which it's transmitted and received. For most purposes, Group 3 machines are now the standard and are likely to serve your needs for many years.

Group 1, found on early fax machines, was the first standard approved by the CCITT. Group 1 transmission is extremely slow. Because the entire technology is outdated, don't even consider purchasing Group 1 machines.

Group 2 also employs an analog method but has faster transmission rates (about three minutes per page). Unless you have clients who use them, you should have no need for Group 2 machines.

In Group 3 technology, an image is scanned and the information is *digitally* encoded. In other words, each pixel, or dot, of the original is assigned a number. That number is then transmitted to the receiving machine, which prints the image that corresponds to that number. The two digital coding methods that are used in Group 3 fax technology are called *read* and *modified Huffman*.

The most important breakthrough for fax technology has been the speed at which information is transmitted. Group 1 machines take six minutes to send or receive a one-page document that a Group 2 machine can handle in about three minutes. Group 3 machines take less than a minute to send the same document, and some take as little as 12 seconds.

Because fax standards for compatibility are agreed upon internationally, all machines within a group can communicate with each other. That means all Group 3 machines can communicate with one another no matter where they are.

But a fax machine cannot communicate between groups unless it supports that group's protocol. For example, a basic Group 3 machine can't communicate with a Group 2 machine. But because Group 3 machines were introduced after there was an installed base of Group 2 machines, some fax manufacturers included both Group 2 and 3 protocols in their machines. This allows a Group 3 sender to transmit to either group of machines.

However, since Group 2 machines are now obsolete, you probably won't want to bother buying a Group 3 machine with Group 2 capacity, unless you know you'll be transmitting to businesses that haven't upgraded to Group 3.

Furthermore, many advanced features found in Group 3 faxes cannot be used when communicating with machines of other groups. For example, Group 2 cannot receive documents with the same resolution and clarity as Group 3.

Group 3 is the standard and will continue to be for at least the next several years.

If you're buying a large machine, however, you may want to invest in a unit that's Group 4-compatible. The new set of CCITT protocols lets Group 4 fax machines take advantage of ISDN telephone lines. If you don't have ISDN telephone lines, Group 4 won't be an option for you.

Despite the fact that all Group 3 machines can communicate with each other, they're not all the same. The CCITT established only minimum standards. Several manufacturers have added higher performance to their machines. For example, to be Group-3 compatible, a machine must be able to transmit information at 2,400 baud or 2,400 bits (1 or 0) of information per second. Some Group 3 machines transmit at 9,600 baud or more. Moreover, although all Group 3 machines use the modified Huffman or read coding methods to encode scanned information, many machines have optional alternative encoding systems as well.

Transmitting at higher baud rates and using alternative coding schemes increase transmission speed, thus lowering phone costs. Some of the features added to the minimum standards authorized by the CCITT are supported by all manufacturers; others are not. (Special features not included in the Group 3 standards are noted throughout this chapter.)

The Group 4 protocols are being developed to work on digital networks. However, ISDN networks won't be significant to most fax users for several years. Indeed, many telecommunications managers joke that ISDN stands for "I still don't need (it)." It's clear that Group 3 fax should serve your purposes for some time to come.

Resolution

As in photography, resolution refers to the detail with which a document or image can be reproduced by a fax machine. Basically, when it receives a document, a fax machine reconstructs the image of that document dot by dot. The more dots per inch, the finer the resolution and the clearer the image.

 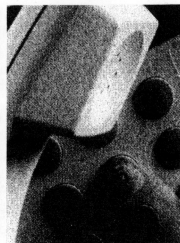

Normal transmission   **Halftone transmission**

The Group 3 standard defines two levels of resolution—
regular and *fine*. The type of resolution you use depends
largely upon the amount of visual detail you need in your
documents. Regular resolution is best suited for text-based
information and simple line drawings. Fine resolution is
most appropriate for material with very small print and
illustrations with fine detail.

Regular resolution constructs images consisting of 203
horizontal lines per inch and 98 vertical lines per inch. That
means each inch of the document is divided into 203 horizontal lines, and each inch of horizontal line consists of 98 dots.
In comparison, most laser printers have 300-dot-per-inch
resolution. A good-quality, 12-inch-diagonal computer display has a resolution of about 100 dots.

This is the sentence in its original form.

This sentence was faxed in regular resolution.

This sentence was faxed in fine resolution.

In fine mode, the number of horizontal lines remains the same, but each inch contains 196 dots instead of 98. Because each fine-resolution dot is half the size of a dot in regular resolution, an image consists of twice as much information and thus has significantly more detail.

Although it delivers more detail than most computer and television screens, fax still cannot produce the fine detail of a laser printer. As laser printers and faxes become integrated (and advanced Group 4 modes become available), fax will gain even higher resolution.

Very small documents that contain a great deal of detail may not be legible when transmitted in regular resolution. On the other hand, fine resolution takes significantly longer to transmit, because it contains more information.

Some manufacturers offer their own nonstandard resolutions to transmit documents that require an even greater degree of detail. However, these resolution schemes may not be compatible with other manufacturers' fax machines, in which case the machines can still communicate using the fine and regular resolutions.

Paper

Fax machines use two types of paper: thermal and regular. The thermal printing process requires thermal paper, which is coated with a colorless dye that responds to heat. Instead of striking the paper like a typewriter, the fax machine's print head heats the paper. As the paper is heated, the dye turns black, forming the image. Of the fax machines currently in use, over 80 percent use thermal paper.

Fax machines that use regular bond paper usually employ the same printing process as photocopy machines. A few use a process called thermal transfer, in which heated pins transfer ink from a special ribbon onto the paper. Another technique is the inkjet process; still others use dot matrix printing, as do many computer printers. However, most fax machines that handle bond paper use the xerographic process found in photocopy machines. Documents received on these machines have the same look and feel as a photocopy.

Each type of paper has advantages and disadvantages. Thermal paper generally costs more. Because it's usually sold in rolls instead of sheets, the recipient must cut the message after receiving it or have a fax machine with an optional cutter feature. Thermal paper looks and feels less professional than normal paper. Finally, thermal paper tends to degrade under high heat. For example, thermal paper left on the dashboard of a car on a hot summer day will turn completely black!

On the other hand, the thermal printing process requires no toner or drum (unlike the photocopy process). Further, a thermal printer has almost no moving parts; consequently, maintenance costs are significantly lower. Because the paper comes in rolls, thermal paper can produce documents up to 39 inches long, a handy feature for faxing spreadsheets, large drawings, architectural plans or other nonstandard documents.

Poor-quality thermal paper can hinder your machine's performance. Not only can it result in light or faded copies, it can leave damaging residue on your machine's thermal print head. Beware of "great deals" on fax paper; consult your dealer or the manufacturer of your machine.

The difference between thermal paper and thermal transfer printing can be confusing. Many thermal transfer machines are advertised as "plain paper" fax machines. However, they don't use the laser or photoelectric processes used with high-end units. Although they cost around $3,000 or less, the thermal ribbons they require are expensive. Consequently, they have a higher cost per copy than thermal paper printing or photoelectric and laser printing.

Bond, or "plain," paper offers two advantages. Documents on bond paper generally look better and resemble papers you work with every day. In addition, the paper costs less.

But bond paper has its limitations. The xerographic process requires that toner be replenished and the print drum be replaced periodically. Most bond paper comes in standard 8 1/2- by 11-inch sheets and can't receive longer documents in one piece. When long documents are received, single sheets must be taped together to record the complete image.

As fax technology is integrated into computer technology, fax documents eventually will be printed by the same printers your computers use. Chapter 6 discusses this and other special features.

Document Size

When evaluating a fax machine, be sure that it can accommodate the page sizes you'll need. An engineer trying to transmit drawing plans has very different needs than an import/export service transmitting short memorandums.

Each fax machine has a minimum and maximum document size it can transmit. The most common minimum is 5 inches by 5.8 inches. Fax machines that have the same type of reader found on photocopy machines (in which you lift a cover and place the document flat on the glass) have no real size minimum. If you plan to send existing business forms via fax, the minimum document specification can be important.

Inexpensive machines designed mainly to send notes cannot accommodate full-size letter paper, but most can transmit a document 8.5 inches wide, the width of a standard sheet of paper. Some, however, cannot transmit information that runs from edge to edge. In other words, the scanner must have margins.

Unattended Operation

Virtually all machines can receive and print a fax automatically without the intervention of an operator.

However, if your machine shares a line with a telephone and doesn't have an auto voice/data switch, the fax machine won't be able to receive messages unattended during those periods of the day when the telephone must be accessible.

Transmission Time

Because it affects your telephone charges for fax transmissions, transmission time (the length of time it takes a document to get from here to there) is a key feature and selling point. It also may affect your administrative costs associated with fax. Consequently, manufacturers often present impressive numbers to prospective buyers to convince them that their machine is the fastest. Unfortunately, the machine's performance in the workplace may not match the manufacturer's performance claim.

Your actual transmission time is influenced by several factors. Consider, for instance, the speed, or baud rate, of the modem (the device that actually sends the information over the telephone lines). As mentioned earlier, Group 3 fax machines send information at a minimum of 2,400 characters per second (or 2,400 baud).

Many machines have built-in modems that transmit at higher speeds—9,600 baud or more—but these high transmission rates can be used only for communicating with fax machines of equal speed. Although 9,600 baud seems to be emerging as the most common transmission rate, some low-cost personal fax machines transmit and receive at only 4,800 baud, or half-speed. Other machines transmit at twice the speed of a 9,600-baud modem. In evaluating a fax machine, find out what baud rates are available for it.

Remember, fax machines must communicate at the same baud rate. In other words, a 9,600-baud machine communicating with a 4,800-baud machine will send at 4,800 baud. Some high-speed modems communicate only with fax machines made by the same manufacturer—and sometimes work only with the identical model. Unless you check these details before buying, you may pay extra for features with limited practical application.

The second factor that influences your transmission time is the resolution of the image to be received. Transmitting and receiving at fine resolution require significantly more time (and telephone cost) than transmitting and receiving at regular resolution.

Several manufacturers include special techniques to speed up the rate of transmission. Some *compress* data, so that only meaningful information is transmitted. Data compression doesn't change what is printed, it merely changes how much information is transmitted over the phone line.

In general, however, compressed data must be *decompressed* at the other end, a feature that may not be available on the receiving machine. In addition, some machines provide for *white space skipping* (i.e., areas of a document that have no image are scanned quickly, speeding transmission time).

Bear in mind that unless both machines support the same method of speeding transmission, the normal standard will be used. As fax becomes more popular and competitive, higher transmission speeds and more efficient transmission features will become the norm. For now, it's important to remember that many of these features must be matched by the features of the receiving machine. Otherwise, a salesperson's promises about high-speed transmission may be technically accurate but of no value in the workplace.

Size and Weight

The earliest fax machines were large, smelly devices, often hidden in the basement. Today some desktop units weigh less than 10 pounds and take up little more desk space than a telephone-answering machine. Some full-featured models with large document feeders, however, can weigh as much as 40 pounds.

Because fax machines work with all telephone lines, some battery-operated models are designed for portability. Indeed, sending faxes from the back seat of a limousine is an emerging status symbol among the rich and frivolous. Auto faxes are becoming popular in the trucking industry, where signed bills of lading are a daily part of doing business. (In fact, many truck stops now offer fax service.)

ADVANCED FEATURES

Advanced features such as the following are available on most models.

1. Transmission features accelerate the actual sending or receiving of a document.

2. Production features improve the quality of the document in hardcopy.

3. Management features contribute to managing fax technology more efficiently.

ADVANCED TRANSMISSION FEATURES

Not surprisingly, because fax is a communications tool—not a production tool—most advanced features are designed to improve and streamline transmission.

Built-in Dial

Although fax machines use regular telephone lines to transmit and receive documents, not all fax machines have telephone-calling capability. With the simplest fax machines, a separate telephone handset must be plugged into the same phone jack as the fax. To send a fax, the user must dial a number by using the telephone and then direct the fax machine to send the document.

Built-in dial capability is perhaps the most commonly available advanced feature. With a built-in dial, a user dials a number directly from the fax machine, eliminating the use of a separate telephone.

Although convenient, a built-in dial cannot be used if a person, not a fax machine, is at the receiving end. Organizations often don't reserve a telephone line specifically for fax usage, because incoming fax traffic may not be heavy enough to justify a dedicated line. Moreover, at some companies all incoming calls may be routed through a switchboard or PBX.

In these situations, called *manual receive*, a person answers the call and then turns on the fax machine. If you frequently send fax messages to companies using manual receive, you must be sure that your machine includes not just a built-in dial but a complete telephone handset integrated into the unit so that you can talk to the person at the other end and ask them to activate their fax machine.

As an economical alternative, you can add this feature to any fax machine you buy simply by plugging an inexpensive telephone into the same telephone jack.

One-Touch Dialing

Fax machines with built-in dialers often offer alternatives that automate and accelerate the speed at which another fax machine can be dialed: *one-touch dialing* and *speed dialing*. As with memory telephones, frequently used telephone numbers can be programmed and stored, then dialed by pushing the button assigned to each. One-touch dialing is a particularly useful feature if you plan to send faxes regularly to a specified group of recipients, such as a small network of field offices or a few major customers.

If your telephone system requires that a number be dialed to access an outside line, the fax machine's one-touch dialing must have the ability to *pause* between digits. Many machines include a pause key or a special character that produces a two-second delay. Consequently, to program the fax machine to dial the number 555-1212 in a telephone network that requires dialing 9 to access an outside line, the user enters 9-pause-5-5-5-1-2-1-2.

In analyzing a fax machine's one-touch capability, be sure that each memory can hold a sufficient number of digits to meet your requirements. Dialing a long-distance number within the US requires at least 11 digits, perhaps more, depending upon the requirements of your telephone system. Dialing an international number requires 14 or more digits.

Speed Dialing

Speed dialing isn't quite as fast as one-touch dialing but is much faster than dialing the complete fax number each time. In essence, speed dialing allows telephone numbers to be stored and accessed using a two-digit code. As a result, you can place a call with four keystrokes: one stroke to access the speed-dialing feature, two strokes to access the number, one stroke to initiate the dialing process.

Many telephone users have found speed dialing a frivolous feature. The time you save by reducing the number of keystrokes is more than offset by the time consumed by programming speed-dialing numbers and then looking them up. However, with fax machines, speed dialing can be a useful technology.

First, many companies these days have separate fax numbers, which few people outside a company bother to memorize. Nor do they dial the number so frequently that they learn it by rote. (Indeed, many executives don't even know their own fax number; when asked, they have to check their business cards!) As a result, most fax operators refer to a directory for each number they must dial. By keeping a directory of your 30, 50 or 100 most commonly used numbers near the fax machine, the speed-dialing feature can be utilized efficiently.

Second, speed dialing helps to avoid misdialing. It's extremely annoying for a person to answer the telephone only to hear a fax machine on the other end; and when dialing internationally, incorrectly dialed numbers are costly.

Most fax machines with speed-dial capability allow a name field to be associated with each stored fax number. For example, you can program the word *lawyer* to represent your attorney's fax number. To dial that number, you would look up the word "lawyer" in your directory and enter the speed-dial number associated with it (e.g., 02).

Some machines display the number dialed and/or the word "lawyer" on an LCD screen. A stored number directory that includes character names can be useful, because it eliminates the possibility of accessing the wrong fax machine—by dialing 02, for example, instead of 03.

Sending sensitive information to the wrong fax machine can be embarrassing, if not dangerous. But entering the names into a stored number directory has several drawbacks. Most fax machines don't have typewriter-like keyboards. Instead, they have numbers that can be used to represent letters. For example, registering the name "First City Bank" would require the following sequence.

```
16-19-29-31-32-10-13-19-32-37-10-12-11-25-22

F  I  R  S  T    C  I  T  Y    B  A  N  K
```

Although not difficult, it's laborious and time-consuming. Alternative methods devised by some fax machine manufacturers for programming names are just as cumbersome.

Moreover, in some fax models, not enough letters can be stored in the directory to spell out the full name of the party to be dialed. For example, First City Bank requires 15 characters of storage (including the spaces), too many for some fax machines.

Dial cards let you store fax numbers on credit-card-size memory cards, which act like separate speed-dial memories. Although they provide a lot of storage for a small machine, they must be located and inserted, and are usually limited to 20 numbers per card. If you call a large number of speed-dial numbers, it's best to store them in your machine. But these cards are useful for storing numbers called occasionally.

Automatic Redialing

Because fax equipment uses standard telephone transmission lines, your machine will get a busy signal if the receiving fax machine is in the process of sending or receiving information.

Automatic redialing instructs your machine to continue trying to reach the same fax number at periodic intervals. With some models, the machine can be programmed to keep trying until it succeeds. Others will stop after redialing two or three times.

The usefulness of this feature depends upon how often you anticipate getting busy signals when accessing other fax machines. Automatic redial lets you leave your machine unmonitored, even if the receiving machine is busy.

Alternate-Number Dialing

As fax technology becomes more common, busy signals will increase. Many companies will have several fax machines, each plugged into its own dedicated line. A fax machine with *alternate-number dialing* can be programmed to dial another telephone number automatically if the original number dialed is engaged.

Most machines with alternate-number dialing also have automatic-number redialing. Thus, if both numbers for the fax machine dials are engaged, it can be programmed to keep dialing—alternating between the two numbers—until one of them connects.

Of course, spending several hundred dollars for automatic redial and alternate-number dialing may not make sense if an operator is monitoring the machine, knows when a receiving machine is busy and can redial later. But if you wish to leave the machine unattended, both automatic redial and alternate-number dialing are important features.

Transmission Reservation

Transmission reservation lets a document be prepared and programmed for transmission while another is being received, a useful feature for busy machines.

Delayed Transmission

Both automatic redial and alternate-number dialing are useful for *delayed transmission*, a feature that programs the fax machine to send a document at a specific time—perhaps when it's unattended by an operator. (See Chapter 3 for a detailed discussion.) In those situations, if the sending machine receives a busy signal, it continues dialing the same number or switches to alternate numbers until the document is successfully sent.

Call Notification

Call notification is convenient if your fax machine is located in an out-of-the-way place. When a message is received, the machine automatically calls another telephone extension and signals that a fax has arrived.

Automatic Voice/Data Switch

In many companies, a fax machine is plugged into the telephone network at a station where a telephone also is used. Indeed, some compact fax machines, complete with integrated telephone handsets and answering machines, can serve as a multifunction communications center.

An *automatic voice/data switch* allows the fax to determine whether the caller at the other end is a person or another fax machine. If the caller is a person, the telephone in the receiving unit rings. If the call is from a fax machine, the unit accepts the fax transmission automatically, without ringing the telephone. For many small businesses that don't want to dedicate a line exclusively to fax use, automatic voice/data switching is essential.

This feature lets you direct a fax transmission to a specific telephone number. For example, if a company has three incoming telephone lines, a fax with a voice/data switch can be plugged into the third line and that number can be publicized as the company's fax number. If a call comes in on that line, the machine automatic with voice/data switch senses whether it's a fax or a telephone call and responds accordingly.

Many units work by listening for a fax calling signal. This isn't the fax modem squeal, but a faint 1100-hz tone heard for a half second every three seconds. If this tone is detected, the call is transferred to the receiving fax machine; if no tone is heard, to the phone. However, fax/phone switches have a limitation you should be aware of. The call must be answered by the fax/phone switch first, to check whether a fax or a person is calling. The caller may not notice this, since many switches send a false ring back that makes the caller think that the phone is ringing at the other end.

Since most fax machines try to connect with another machine for up to 30 seconds after a connection is made, the integrated fax/telephone answering machine plays a short outgoing message for 12 to 16 seconds, then listens for a voice for several seconds. If it doesn't detect one, the machine switches to fax mode. The unit can also both record a voice message and handle a fax transmission from a single call. If the call was initiated by a person wanting to send a fax, the outgoing message instructs the caller to press the * key on the telephone, which activates the fax unit.

If your fax machine shares a line with other telephone equipment, be sure to look beyond the manufacturer's claims and determine exactly how the switching works. The right implementation can solve your problem; other implementations will just make it worse.

In assessing the automatic voice/data switch feature, it's extremely important to check with your telephone equipment supplier to make sure the fax machine is compatible with your telephone system. (For more information, see Chapter 4.)

This feature is also available as a separate device that connects both your telephone and fax into the telephone line. Available for around $200, this add-on device may cost you less than buying a fax that includes the feature.

Document Feeder

Although rather low-tech, the *document feeder* is one of a fax machine's most useful features. The document feeder inserts the original document into the machine one page at a time and eliminates the need for manual insertion of each sheet. Although having a document feeder doesn't increase the number of pages that can be sent, the disadvantages of manual-feeding long documents are obvious.

Document feeders are essential for delayed transmission (see Chapter 3) so that the document to be sent can be placed in the document feeder to await the specified transmission time. In this situation (as in the case of unattended transmission), the capacity of the document feeder will determine the utility of the fax machine. Document-feeder size also affects the price of a fax machine.

Most commonly, document feeders can hold 5, 10 or 30 sheets of paper. A five-page document feeder restricts you to sending a maximum of two faxes at a time, because even a one-page fax takes two pages (a transmittal sheet and the sheet with the message itself).

Instead of using a document feeder, some fax models work more like copiers. The original is placed on a flat piece of glass and scanned. While this method is as inconvenient for sending long documents as it is for photocopying them, it's the only way to fax documents that are too cumbersome for a document feeder, such as pages from a book. A library, for example, would be well advised to consider such a model.

Automatic Paper-Cutting

Many fax machines use paper that comes in rolls rather than sheets. This allows the machine to receive documents of a specific width but of variable length. However, when a multipage document is received, pages are printed one after the other on a continuous length of paper.

With *automatic paper-cutting*, each page is automatically cut from the roll after it's received. Pages of variable length can still be received, but each page is physically separated from the next. This feature, most commonly found on more expensive models, eliminates the annoying necessity of cutting paper manually.

Copier Option

Because fax machines reproduce original documents, many manufacturers have begun to include a *copier function*. In copier mode, the fax machine operates like a simple office copier, making photocopies in all the resolutions the machine can support.

For small businesses and individuals who make only a few copies, a copier mode on the fax machine can be an attractive alternative to investing in a separate copier. However, there are limitations. Unlike copiers, fax machines can make only one copy at a time. Consequently, to make five copies of a document, the original must be fed through the machine five times. (The exception is a machine with internal memory.)

Moreover, it takes a fax machine about 30 seconds to reproduce an original, which is extremely slow. And the paper needed for fax copies costs substantially more than regular bond copy paper.

Nevertheless, the copier function can be an attractive feature, even if you don't need office copies. It offers you a rough preview of how a document will look when received. This is particularly useful if you send complex documents with many details. Also, the copier function lets you experiment with fine mode, regular mode and halftone, and then decide which is the most appropriate prior to actual transmission.

Fax technology quickly is becoming integrated with copier technology in larger multipurpose machines designed to serve as the primary copiers in the office. With integrated fax/copiers, your copying and fax needs must be assessed independently of each other, and the machine must be able to meet both requirements.

Automatic Fallback

As mentioned earlier, fax machines send information over the telephones via modems, which send information at different rates. Most fax machines send information at high speeds (baud rates). Unfortunately, the faster the information is sent, the more likely that errors in transmission will occur and that some information will get lost or garbled due to "noise" on the telephone lines. As a result, portions of the document may be unreadable when received.

Fax machines with *automatic fallback* can detect the occurrence of errors and will begin to transmit at a slower rate. For example, many Group 3 fax machines transmit at 9,600 baud. If a machine with automatic fallback detects errors, the machine shifts to 4,800 baud or even 2,400 baud. Though the transmission time is longer, the document received will be more readable.

Automatic fallback is particularly important for transmission overseas and to remote places. If you cannot hear somebody on the other end of a telephone line because of "dirty" transmission, chances are good that the fax machine cannot "hear" either, and automatic fallback may prove to be a valuable feature.

Transmit-Terminal Identification

In essence, *transmit-terminal identification* stamps each page of a fax document as it's sent with your fax telephone number, including the date and time of transmission and the page number. Some fax machines also can insert phrases— such as the name of your company—in the heading. This identification is useful for fax transmissions that are several pages long. However, all transmissions should be accompanied by a cover sheet that includes the information listed

MODE	CONNECTION TEL	CONNECTION ID	START TIME	USAGE T.	PAGES
TX	16883909	CH PRINTING	08/30 12:37	00'49	01(00)

RECEIVE REPORT

JUN-26-88 SUN 8:44

TERMINAL #

#	DATE	S. T.	NAME	TIME	PGS	NOTE	DP
01	JUN-26	8:37	0	2' 6"	1	OK	
TOTAL				2M 6S	1		
GRAND TOTAL				2M 6S	1		

above and the number of pages in the document.

Many fax machines stack incoming messages in order, face up. Therefore, the last page transmitted is on the top of the pile. Individual page identification helps the receiver collate the incoming document. Moreover, having page numbers that correspond to the fax document, and not necessarily the original, can be extremely helpful, particularly if you're sending only part of a document that isn't numbered in consecutive order.

Transmit-terminal identification should be viewed as a convenience for the receiving party and insurance that your document will arrive intact at its destination, particularly if it must pass through several hands at the receiving end before it reaches the person it's addressed to.

Batch Index

The *batch index* places a mark on the top of each page for each incoming fax. After a fax is received, the mark moves a little to the right for the following page, and so on. Because each page in a multipage fax will match the marks of other pages in that batch, a recipient can easily find the start of one document from the end of another. It makes separating many pages of faxes easier each morning.

Polling

One of the most misunderstood fax features, *polling* allows a fax machine to call another to request a document rather than send one. Polling also lets you place documents in the fax machine that will be sent only at the request of a calling machine.

For example, a supplier of fine china has 100 independent salespeople selling its wares door to door. Each salesperson has a fax machine, and at the end of each workday, the manufacturer receives a sales report.

Without polling, each of the salespeople would have to call the supplier's fax machine. With 100 incoming calls beginning after the workday ends, the entire activity would have to be well-coordinated or it would take several frustrating hours to complete. With polling, the salespeople place their sales reports in their fax machines but don't send them. Instead, the manufacturer's fax machine automatically calls each machine in order and requests the sales report. With automatic speed dialing, the procedure can be quick and efficient and require no operator intervention.

Both the transmitting and receiving machines must be prepared in advance to poll (request documents from others) or to be polled (send documents on demand). To poll, a person simply loads a document into the fax machine and presses the poll function. In its basic form, any machine can call the machine that has been set up to poll and request the document. This is called *free polling*.

In some situations, however, free polling can be dangerous. Suppose, for example, that a builder wants to poll several potential vendors for quotes on building materials. With free polling, an unethical competitor could simply request a competitor's quote. Indeed, with free polling, the party to be polled can't control who receives the information, which brings up the subject of security.

Security

With the above situation in mind, several manufacturers include various levels of *security* with their polling function. The most rudimentary secure polling procedure requires that you program a list of the telephone numbers for all fax machines that have polling privileges. Although this system limits the number of outsiders who can poll the machine, it doesn't ensure that the information left in the fax machine to be polled won't be picked up inadvertently by an inappropriate recipient with polling privilege.

A second level of security requires not only that the telephone number of the machine requesting the document be listed, but that the terminal identification number match before the document is sent. This ensures that only authorized machines can poll your machine for a specific document.

For the most advanced level of security, not only must the telephone number and the terminal identification number be known, but the polling machine also must transmit a special security code. Consequently, the number of people who can poll your machine is limited. Unfortunately, this level of security generally works only with machines built by the same manufacturer. So, if you plan to send confidential information, you may be restricted to poll only with equipment made by the same manufacturer.

If you plan to gather information routinely from a large number of sources, polling is an essential feature. In such situations, polling manages the flow of fax traffic. Because the machine can contact each source of information, the need for human intervention is virtually eliminated.

Delayed Transmission/Timer Transmission

To take advantage of lower telephone rates or big time-zone spreads, it's often more convenient to send a fax transmission at night. *Delayed transmission/timer transmission* allows documents to be sent to specific fax machines at specific times. Delayed transmission is usually limited to machines with either document feeders or internal storage memory. Delayed transmission offers perhaps the greatest potential for cutting fax costs. In practice, however, it's often ignored because it involves mastering a complex procedure.

Despite the learning curve involved in effectively using delayed and timer transmission, it may be worth the investment. For example, AT&T rates for transmitting from Los Angeles to New York drop by about 50 percent after 11 PM. If a five-page report is to be sent from Los Angeles to New York every day, delayed transmission can save you hundreds of dollars a year.

However, the usefulness of the delayed transmission feature depends on several factors. First, only documents that can be stored in the sheet feeder or in the internal memory can be transmitted on a delayed basis. If a wealth of material must be transmitted every night, a machine with an internal memory is probably your best option.

Second, most machines can be programmed to send only one delayed transmission to one location at a time. If you plan to send material to several locations, you must be certain that the machine has that capability.

Programming for delayed transmission to different locations can be complex. Either you must enter the exact number of pages of each document to be sent to each location, or you must prepare specially coded sheets, called *optical mark reader sheets* (see below), to signal the machine when one document has ended and another is beginning. If your fax traffic is high, delayed transmission can keep your fax working 24 hours a day and is worth investigating.

Optical Mark Reader Sheets

Optical mark reader (OMR) sheets are most familiar in the form of the answer sheets for standardized tests on which you fill in the appropriate circle with a No. 2 pencil. Some fax machines can read OMR sheets that contain information directing them to send a specific document to a specific location at a specific time.

In general, OMR sheets are used for unattended broadcast transmissions and delayed broadcasts. Each document to be sent is loaded into the document feeder with an OMR sheet detailing the instructions for the fax machine. Most machines require that the destination number already be stored in memory. The machine waits until the time specified on the OMR sheet, then sends the document.

OMR sheets aren't the only way to program unattended transmissions of multiple documents. Some machines can be programmed through their control panel in much the same way that a VCR can be programmed to record television programs at different times.

Memory Features

Perhaps one of the most exciting recent trends in fax technology is the emergence of less expensive equipment with lots of document memory, which electronically stores scanned material. Generally, memory capacity is measured by the number of pages a machine can hold, which can run from 3 to over 100 pages, but photographs or complex graphics can fill several pages of document memory. For people familiar with computer terminology, 15 pages of document memory equals about 256 KB, or 60 pages of document memory per megabyte.

Adequate memory sets the stage for several advanced sending and receiving features.

Broadcast Transmission

Broadcast transmission uses a combination of internal memory and stored fax numbers to "broadcast," or transmit, a document stored in memory to specified fax numbers. For example, price updates can be broadcast to hundreds of sales offices or every weekend after a single copy of the updates is scanned into the central machine. Then, each number on the broadcast list is called and sent the information.

Without this feature, a separate copy of the document would have to be loaded into the document feeder for each location, an approach that's affected by document-feeder limitations. And, you would have to call each receiving number. If your machine serves as a central hub for information, this is a useful feature.

Many manufacturers claim broadcast capability even on their low-end machines. Reading the fine print, however, reveals that they have that capability only when used in conjunction with an expensive machine made by the same manufacturer. Using that method, a document is scanned into the inexpensive machine, and then is sent to the more expensive "hub" machine for true rebroadcast.

If you need broadcast capability, be sure the machine you buy can broadcast directly. Without its own internal memory, it can't perform the broadcast function.

Store and Forward

Store and forward combines memory capability with broadcasting. A machine can be programmed to receive a document, store it in memory and then retransmit it to other machines. Store and forward can help reduce telephone charges, particularly for companies with large fax networks spread across the country or the globe. For example, the headquarters of a large corporation can broadcast the details of a personnel change to several regional headquarters, which in turn rebroadcast the same information to the local branches. Or, a factory located in Asia can send its production figures for the day to a single US location, which rebroadcasts them to regional centers.

Substitute or Backup Reception

This feature stores incoming fax messages in memory if the paper runs out or jams.

Memory Transmission

Memory transmission lets you scan and transmit a document more quickly than if you wanted to scan a document to send directly as with a nonmemory machine. You don't have to hover around the machine waiting for a transmission to be completed. You can also program your machine for delayed transmission and know that the transmission won't be interrupted by a document jam.

Fax forwarding is a version of store and forward. It lets a fax machine store an incoming document in memory and re-send it on demand to another fax machine. Fax messages can automatically follow travellers from place to place.

Memory can also enhance other features. For example, confidential reception is available without memory; however, the recipient must be present to receive the fax. Memory allows messages to be stored in electronic mailboxes; recipients then can collect messages when they're ready.

Delayed transmission also works better with memory. Without it, you need OMR sheets to send delayed transmission to several different locations. Memory allows machines to automatically send documents to several locations at different times of day.

ADVANCED PRODUCTION FEATURES

Advanced production features are designed to enhance the appearance of the document so that the recipient's copy is as close to the original as possible. As with photocopy technology several years ago, the quality of the copy received is one of the most widely discussed issues in fax technology. Not surprisingly, production quality has become a key selling point in the fax industry.

Error Correction Mode

The CCITT optional Error Correction Mode (ECM) can virtually eliminate transmission errors when used with another ECM machine. A small block of the document being sent is placed in an ECM memory area. If the receiving machine supports ECM, and the data are garbled during transmission, that block is then retransmitted from the memory buffer before it's printed on the receiving side. Although this only takes a small amount of memory, you will need memory for this feature to work. Machines with internal document memory don't necessarily have ECM, and vice versa.

Image Enhancement Systems

Image enhancement systems are the latest feature to confuse the fax buyer. These image-processing techniques eliminate the jagged look of lines on fax documents. They can work on either the sending or receiving side. More manufacturers are making those systems available, and not only on their high-end models. Even low-cost machines that contain a small amount of memory have this feature.

Contrast Control

In many of its operations, a fax machine is like a copier machine; after all, it makes copies of an original. *Contrast control* helps the machine improve the quality of the copy when the original document is either too light or too dark.

Fax machines are available with either manual or automatic contrast controls. With manual contrast control, if the original document is too light or dark or has too little contrast, you simply adjust the contrast control accordingly. Most models carry only three contrast settings: light, dark and regular.

Machines with automatic contrast control sense the quality of the original and make the necessary adjustments. If you frequently fax complicated, multicolored documents, this feature can be particularly useful.

You can check the quality of the contrast control by operating the fax in copier mode. Experiment with high, low and regular contrast. Keep in mind, however, that telephone line noise often erodes legibility, so the results from the copier will be a "best-case" preview of how the fax will appear on the receiver's end.

Oversized Document Reduction

Those who plan to send spreadsheets printed on 11- by 14 7/8-inch computer paper or an engineering drawing on 11- by 17-inch paper will be particularly interested in reducing such oversized documents. The scanner size determines the largest document that can be reduced, usually 11 inches wide. The *oversized document reduction* feature allows large documents to be received on standard-sized paper with all the information intact, albeit smaller.

Fax machines won't print out large documents; they print them in the reduced size. So, if you own a reduction copier and don't need to send a large volume of oversized documents, this feature isn't necessary.

On the other hand, if you work regularly with oversized documents, this feature may be more convenient than the extra reduction step.

Halftone Transmission

Most black and white photographs are made up of various shades of gray, or halftones. *Halftone transmission* lets a fax machine record and reproduce subtly shaded photographs and graphics more accurately.

Halftone transmission is a feature determined almost exclusively by the transmitting machine. That is, if your machine has halftone capability, it can send halftone images to any Group 3 machine. The only exceptions are receiving machines with paper too insensitive to record the more subtle signals.

Halftone transmission generally is rated according to the number of different shades of gray it can transmit. Fax machines are available with 8, 16 and even 64 shades of gray.

If you intend to transmit a high volume of photographs and graphic material, test the halftone transmission capabilities *over real telephone lines*. The copier function can't provide a realistic test of the machine's capability. Because halftone transmission entails sending more complex information, the transmission is more susceptible to telephone line noise and other interference.

RS 232 Port or Interface

Familiar to computer users, the *RS 232 port* is essentially a plug that allows two machines to be joined by a cable and pass information back and forth.

An RS 232 port on a fax machine opens up several options. First, it allows the fax machine to serve as a printer for a computer. Unfortunately, because most fax machines use thermal paper, they're much more expensive to operate than a standard computer printer—and the quality of the work is lower. Furthermore, most fax machines won't print graphic images sent by a computer. Finally, an RS 232 port probably will add more to the cost of the fax machine than the price of a good dot matrix printer.

Some newer models allow data to be sent from a computer to a fax machine using the RS 232 port. Other models enable the fax to serve as a scanner for a personal computer. Study the exact capabilities of each machine carefully to ensure it will serve your needs.

ADVANCED MANAGEMENT FEATURES

The final step in evaluating a fax machine—one that can prove costly in the long run if ignored—is the ease with which its use can be monitored and controlled. Reporting features are important to ensure efficient fax management.

Transmission Report

After a fax is sent, a *transmission report* records the telephone number of the receiving machine (and sometimes its terminal identification number, too), the number of pages sent, the transmission time and an indication of any errors in transmission. Attaching a transmission report to the original document is an efficient way of recording who has received a document and when.

Some expensive fax machines stamp the transmission report on the original, but this can be confusing if the document has to be retransmitted elsewhere at a later date.

Communications Journal

The *communications journal* provides an audit trail of all fax activity. It lists the date, time, duration, length and destination or source of all fax transmissions and receipts. The report can be printed on demand or after a full page of information has been accumulated.

The communications journal is a key management tool, indicating whether the fax machine is being used according to established company guidelines and whether faxes are being sent at the most economical times and to the right sources. It offers a cross-check for any fax logs you might maintain and a safeguard against unauthorized use.

Finally, for those companies that bill clients for the faxes they send, the communications journal is a useful aid in preparing invoices.

```
*************************************************************
*                                                    P.01  *
*                       SEND  REPORT                       *
* TERMINAL #   001                     JUN-25-88 SAT 16:49 *
*                                                          *
* NO. DATE  START       RECEIVER   TX TIME  PAGES  NOTE    *
*                                                          *
*  1 MAY-27  6:47 3518256             48"     1     OK     *
*  2         7:12 12134704790       4'35"     2     OK     *
*  3 MAY-31 17:41 3518256             58"     1     OK     *
*  4 JUN-16 16:19 3518256           2'50"     1     OK     *
*  5 JUN-18 11:17 3518256             52"     1     OK     *
*  6        11:22 3518256             47"     1     OK     *
*  7 JUN-21 16:00 3518256             55"     1     OK     *
*  8 JUN-25 16:35 3519698           1'08"     1     OK     *
*  9        16:36 3519698           2'08"     1     OK     *
* 10        16:40 3519698           2'08"     1     OK     *
*                                                          *
*                    TOTAL        17'09"    11             *
*                                                          *
*           GRAND TOTAL  TIME:   36M 48S  PAGES:    22     *
*                                                          *
*************************************************************
- - - - - - - - - - - - - - - - - - - - - - - - - - - - - -
*************************************************************
*                                                    P.02  *
*                     RECEIVE REPORT                       *
* TERMINAL #   001                     JUN-25-88 SAT 16:49 *
*                                                          *
* NO. DATE  START       SENDER     RX TIME  PAGES  NOTE    *
*                                                          *
*  1 MAY-27  6:49 3518256           1'21"     1     OK     *
*  2         9:30 3518256             52"     1     OK     *
*  3 MAY-28  6:29 3518256             52"     1     OK     *
*  4 MAY-29  3:49                  **'**"     0     CANCEL *
*  5 JUN- 6 11:46                  **'**"     0     CANCEL *
*  6 JUN-16 15:18                     19"     0     CANCEL *
*                                                          *
*                    TOTAL         3'24"     3             *
*                                                          *
*           GRAND TOTAL  TIME:   11M 27S  PAGES:    10     *
*                                                          *
*************************************************************
```

Moving On

Although advanced features are packaged in various combinations, they can be effectively analyzed in three areas: transmission features, production features and management features. The advantages of each group of features depend entirely on how you plan to use fax. In some situations, the halftone transmission feature is superfluous; in others, it's essential. Your goal, and the goal of the next chapter, is to match features to your needs and budget.

CHAPTER 3

MATCHING FAX TO YOUR NEEDS

CHOOSING THE RIGHT FAX

When selecting a fax machine, be ever-mindful of the age-old axiom, "Get the right tool for the right job." Chapter 1 clarified the ways in which fax communication can be used. Chapter 2 introduced you to the major features available today. This chapter helps you select a model that not only fits your budget but is equipped with the features that will meet your needs.

Determining which fax is best for you involves three steps:

1. Review the features you need.

2. Determine which combination of features (and prices) are available. For example, if you wanted an automobile with a leather interior, you might have to buy the mag wheels as well. So, too, in fax technology, features come in clusters that command different price levels.

3. Buy the machine from the appropriate vendor.

As you evaluate equipment, you may want to use the following checklist to compare each machine.

COMPARING FAXES			
MODEL			
Price			
Modem Speed			
Transmission Speed			
Group Compatibility			
# of One-touch Autodials			
# of Speed Dials			
Auto Redial			
Delayed Dialing			
Halftone Transmission			
Fine Resolution			
Superfine Resolution			
Contrast Adjustment			
Auto-document Feed (# of pages)			
Security			
Reduction			
Auto Reduction			
Scanning Width			
Paper Cutter			
Polling Locations			
Polling Turnaround			
Polling Sequential			
Polling Secure			
Memory (# of pages)			
Error Correction Mode			
Image Enhancement			
Broadcast			
Paper Length			
RS 232			
Transmission Confirmation			
Reception Advice Report			
Activity Report			
Dimensions			
Weight			
Comments			

PRICE VS. FUNCTION

In the long run, the most important consideration in choosing a fax machine is function—how will the machine you buy enhance your business communications? But prices vary widely, so cost considerations can't be overlooked. In fact, price is perhaps more important in choosing fax equipment than in choosing other kinds of office machines. If you select a fax that pays for itself quickly, you haven't made a mistake. But you may save more money by buying a fax machine with additional features that further enhance your communications.

However, cost isn't the only criterion for evaluating a fax machine. Because Group 3 compatibility promises to be the most important standard for several years, the fax machine you buy today will remain useful for years to come.

This chapter follows the three-step selection process outlined above. It begins with a review of the major fax features. A checklist is provided so that you can more easily determine which features best suit your needs.

Following the review is a general analysis of what features are available at various price points. As with most new technologies, fax machine prices are plummeting, and new models with new combinations of features are being introduced. It's important to check the market for the latest advances and prices.

Finally, when you know which features you need and what price you want to pay, you must decide from whom to buy. Fax machines are sold through various distribution channels, from retail stores to mail order companies. The benefits and drawbacks of each distribution channel are addressed in the final section of this chapter.

Instead of assessing features individually, try to determine which features work together to provide additional benefits. In a way, a fax machine is like a microwave oven—it works just fast enough to make you want to stand and wait until it finishes the job. But if you buy a machine with adequate memory, automatic redial and alternate number dialing, you can encourage people to load their documents, start their transmissions and move on.

ASSESSING THE FEATURES

Compatibility

Compatibility is perhaps the easiest feature to assess. Don't settle for anything less than Group 3 compatibility. You may see small advertisements in your daily newspaper touting faxes "new out of the box" for less than $400. Be careful. Those machines often are only Group-2 compatible. And although many Group 3 machines can communicate with them, Group 2 machines are awkward, slow and expensive to use.

Bear in mind that Group 2 machines require three minutes to send or receive *a single page*. Hence, they tie up the telephone for long, costly periods of time.

Group 3 compatibility is essential. However, you also may need Group 2 capability if any of the people with whom you plan to communicate have only Group 2 machines. If you expect to be a heavy fax user, survey the people who you'll be faxing documents to. If all have Group 3 machines, Group 2 compatibility is superfluous. If some have only Group 2 compatibility, then your machine also should include Group 2 compatibility.

Group 1 machines are by and large obsolete, and you shouldn't consider paying a premium to add Group 1 capability to your unit. Group 1 requires six minutes to send or receive a page. So, if you have an important client who has only a Group 1 machine, it might pay to buy him or her an inexpensive Group 3 machine. You probably will recoup your costs in lower telephone charges and administrative overhead, and your customer is sure to be grateful.

Notes:

_____ Number of Group 2 Receivers (if any)

_____ Number of Group 3 Receivers

Receiving Paper Capacity

Often overlooked, a fax machine's paper capacity is one of the most important keys to its utility.

If you generate a lot of fax traffic, paper capacity determines how long the machine can receive messages without operator intervention.

For example, many law offices have begun to fax all but their most sensitive documents. A typical legal brief easily can be 50 or more legal-size pages. If the receiving fax machine has a 30-meter (98-foot) paper roll, it can receive only two such documents unattended.

There are three standard sizes for paper capacity.

Roll Size	Approximate Number of Standard Sheets
30 meters (98 feet)	90
50 meters (162 feet)	150
100 meters (393 feet)	300

If you expect to receive a lot of faxes or several large documents at once, large paper capacity is obviously essential.

Several factors can reduce the amount of paper available for actual fax messages. Fax protocol dictates that every fax message be sent with a separate transmittal sheet. Most machines send one or more lines of terminal identification information per page, adding slightly to the amount of paper needed to print a page. Many machines sending documents also transmit a reception report, printed out after each fax is received, adding a few more inches of paper to the fax. Receiving machines also print a reception report when a fax message is completed.

Many manufacturers signal that the paper roll is nearing the end by placing a red stripe down one side of the paper. The unsightly stripe can run for several feet. If you opt to change rolls as soon as the stripe appears, you further reduce the usable amount of the paper.

Clearly, attention to paper capacity is critical. Do you need large paper rolls or paper bin capacity? Perhaps you don't. If your machine will be attended most of the time, or you plan to send faxes rather than receive them, paper capacity may not have much bearing on your choice. For unattended machines or ones that will be receiving a lot of faxes, a large paper capacity may be a necessity.

Automatic Paper Cutter

Midrange and higher-end fax machines generally have automatic paper cutters, which cut each page to a standard length as faxes are received. Without the automatic paper cutter feature, an entire fax message would be printed on one long sheet of paper.

An automatic paper cutter is crucial if you receive many faxes from many different people when your machine is unattended, or if you receive long fax mesages. In either case, it's tedious to have to hand-cut messages. As a rule, the greater the human intervention, the greater the possibility for error.

Document Feeder Size

As with computer printers, document feeders can add significantly to the cost of the machine. Consequently, you should decide whether you need a document feeder at all.

Infrequent users may be more willing to feed each sheet into the machine manually. Moreover, if the fax machine will be used primarily as a messenger service, shuttling short notes back and forth at a moment's notice, a document feeder may be superfluous.

A document feeder is essential if you plan to send multiple-page documents or send documents unattended. In either case, consider the number of documents you plan to send.

A 10- or 12-page report requires several minutes to fax. And time never seems to pass more slowly than when you're standing around watching a fax machine work. As with photocopying, if you find yourself daydreaming and don't feed the documents smoothly, you can make costly mistakes.

Sheet feeders and internal memory are required to utilize delayed transmission and polling. Both features, in the long run, can generate significant savings in telephone costs.

Sheet feeders commonly come with a capacity for 5, 10 or 30 pages. Be sure the feeder can handle your normal daily volume. If you plan to send only three faxes of five pages using delayed transmission, your sheet feeder must have a minimum capacity for 15 pages. And it's wise to plan for growth as well.

Notes:

Plan to use delayed transmission or polling?
Yes____ No____

Number of pages of average fax message _____

Approximate number of pages received per day _____

Document feeder capacity needed _____

Telephone Handset

You'll need a handset next to your machine to transmit to people who don't have a dedicated line or to PBX systems. You may also want to consider a fax machine that has a speaker-phone built in.

Document Sending Size

The document width capacity of your fax machine determines the form of information you can send. For example, if you plan to send computer printouts, your fax machine must have either a scanner that can handle documents 11 inches wide or a separate reduction copier.

Scanners that can accept wide documents reduce them to an 8 1/2-inch width before the machine sends them. Documents emerge at the receiving end in standard 8 1/2-inch-width form. Therefore, if you have easy access to a reduction copier—which may reduce the original image more crisply than a fax machine—you may not need the added scanner capability, particularly if you don't foresee sending large quantities of oversized documents.

Take time to evaluate each machine's scanner performance. Some fax machines don't scan the entire width of the page. Like copy machines, they have dead zones at the margins. Be sure to test a machine's scanning ability in the copier mode.

Copier Option

Many people who don't own a separate copier are interested in the fax copier option. The copier option serves two purposes: it lets you preview a fax transmission (showing approximately what it will look like at the other end), and it can serve as a low-volume copier in its own right.

Each use must be evaluated separately. The more complicated the transmission (such as halftones and other visuals), the less accurate the copier preview of the transmission will be, particularly with fast transmission rates. As mentioned earlier, telephone line noise affects the accuracy and quality of fax messages, and the copier option doesn't account for that. On the other hand, it's useful to have a rough idea beforehand of what the receiver will get, particularly with graphic information.

The quality of a fax machine copy generally isn't as good as one made on a regular copier. Thermal paper, the kind most commonly used by fax machines, has an annoying tendency to turn black under heat; consequently, it doesn't store well. Although fax can double as a photocopy machine in a pinch, if you need photocopying capability regularly, the most prudent course is to buy a photocopy machine.

Resolution

Even the least expensive Group 3-compatible machines support both regular and fine mode resolution. However, if you plan to send detailed graphics or small text, you may require even higher resolution. Unfortunately, the ultra-high resolution modes aren't standard among manufacturers. And you often can send such images only to machines built by the same company.

Halftone Transmission

If you plan to send photographs or drawings with tones or screens, you must have the halftone transmission feature. If you'll be sending only black-and-white text and graphics (line drawings), you won't need it. All Group 3 machines can *receive* halftone transmission, regardless of whether they have that feature.

Notes:

Do I need to send photographs or grayscale graphics?

Yes_____ No_____

Superfast Mode

Generally, the faster a message can be transmitted, the more cost-efficient the fax communication. Some fax machines come equipped with their own methods for speeding transmission by compressing the data. White-space skipping is an example. When the scanner encounters white space, rather than send several blank lines, it sends the receiving machine a message to skip the entire line—a kind of electronic shorthand.

Unfortunately, if the receiving machine doesn't understand shorthand, you can't communicate in superfast mode. Indeed, the highest speeds can be used only by faxes that use the same shorthand.

Although different manufacturers may use the same shorthand protocols for high-speed transmissions, machines made by the same manufacturer often enjoy an added advantage of being able to communicate with each other even more rapidly.

The way machines determine how to communicate with each other is called *handshaking*. Machines built by the same manufacturer quickly ask each other whether they speak the same high-speed language. Other machines may take longer to find the fastest mutually acceptable code to send and receive information.

High-speed capability saves both time and money. It can be an important feature if you have a large volume of international fax traffic and/or if you'll be communicating with machines that use the same high-speed shorthand. Conversely, if your fax communications are mostly local, the extra time you save may not merit the extra cost involved.

Note that actual transmission time may not always meet the theoretical performance specifications. Consequently, if superfast transmission is important, be sure to test the actual capability in a "real-time" situation. Ask the salesperson to send a five-page document to another machine built by the same manufacturer. You may find that a machine advertised to send a page in 20 seconds needs three or more minutes to transmit a five-page document.

Delayed Transmission

By allowing messages to be sent at night, during low-cost calling hours, delayed transmission can cut telephone costs. It also can spread the demand on the fax network more evenly over a 24-hour period.

However, delayed transmission must be programmed, and if only a few fax messages are sent every day or week, it may be more convenient—if slightly more expensive—to send them during regular business hours.

Delayed-transmission capability has the greatest impact when faxing internationally. Often international rates are at their peak during your workday, even though people in another country probably are at home sleeping. Delaying transmission until international rates are lower can produce substantial savings.

The usefulness of delayed transmission is limited by the capacity of your document feeder. Each document to be faxed must be stored in the document feeder (or the internal memory). Consequently, extensive delayed transmission capability requires a full-featured, high-end fax machine.

Dialing Capabilities

Advanced dialing capabilities (one-touch dialing, speed dialing, alternate-number dialing and stored-number memory) can save time and money.

One-touch dialing is most effective if you call a number of locations regularly, perhaps daily or even hourly. Speed dialing, which requires a few extra keystrokes, automates access to a larger universe of fax numbers. Alternate-number dialing lets you automatically call fax machines that are frequently in use without getting a busy signal. Stored-number memory helps avoid misdialing other fax machines.

Perhaps the only drawback to these advanced dialing features is that they require some programming. As many telephone users learned to their dismay about speed dialing, for example, you may find yourself investing in a feature that no one will use.

Store and Forward

Unless you serve as the middle link in a communication network, store and forward probably isn't necessary. Typically, store and forward is used to take advantage of low local telephone rates. For example, if corporate headquarters in New York faxes new pricing information to the Atlanta regional office, that office then can send the information to local sales personnel.

Store and forward also is convenient if you receive information from one source that simply must be passed to others. For instance, if you wish to inform several offices of an individual's schedule and you're using a mobile fax machine, you may want to send that information to another fax, which then will forward it to several local points.

Broadcast Capability

Assessing the need for broadcast capability is extremely tricky, because for many companies it adds a new communications ability rather than replacing or simplifying an existing one.

Broadcasting allows a document to be scanned into a fax machine once and then sent automatically to several preprogrammed telephone numbers. Some machines can preprogram as many as a hundred fax numbers for broadcast. Although many fax machines without the broadcast feature can store a hundred telephone numbers in their memories, these machines must make multiple copies to send your message to the locations. With broadcasting, the message only has to be scanned once.

How can broadcasting be used? A Chicago manufacturer wants to inform distributors and other sales personnel throughout the country about timely price and product changes. Every Monday morning, updated information is faxed to a list of preprogrammed fax numbers. Salespeople have important new data with which to work, and management knows they received it on time without having incurred the considerable expense of overnight mail services.

To decide whether you need broadcasting, identify what you do now that can be automated through broadcasting and decide how it can improve your communications.

Polling

In many ways, polling is the opposite of broadcasting, allowing you to receive information from many locations conveniently. Rather than have each sending machine call the receiving machine—often encountering busy signals— the receiving machine calls and polls the sending machines.

Polling is useful if you plan to do two-way communication via fax. One machine calls another, sends information and then polls the machine to check whether information is waiting for it. This is called *turn-around polling*—a single call carries two messages.

Automatic Voice/Data Switch

Automatic voice/data switch is one of the most complicated and confusing fax features. In theory, it allows a regular telephone user to share a telephone line with a fax machine. If a call comes in for a person, the phone rings. If a fax transmission comes in, the machine automatically accepts it without bothering the person.

This is a popular feature for people who want to receive faxes without operator intervention but whose fax volume cannot justify the monthly charge for an additional phone line. Professionals who spend a lot of time talking on the telephone and perhaps have two incoming phone lines fall into this category, as do field sales personnel and others who work abroad or in different time zones. For example, a company with a European sales representative who works from home can use an automatic voice/data switch to avoid an annoying wake-up call in Belgium at 2 AM.

Unfortunately, different manufacturers have different definitions for the switch feature. In one model, for instance, when the switch detects that the transmission is a fax, employees are greeted by the ear-splitting sound of the fax machine responding in machine language. Only seconds later, if no response is received, does it switch to voice transmission. By that time the caller has hung up.

Heavy fax users should consider a telephone line dedicated to fax. With a dedicated line, people calling your phone number will never get the high-pitched whine of a fax machine. Depending on where you live, a dedicated line costs about $300 the first year and $150 a year thereafter.

Light fax users who plan to be present when faxes are received may also find the automatic voice/data switch unnecessary. The sender can first call to alert the receiver that a fax message is being sent. The receiver can then activate the fax machine.

Security and Secure Mailboxes

Your need for security depends on three factors: (1) the sensitivity of the information you send and receive; (2) the people who legitimately have access to faxed information on both the sending and receiving ends; and (3) the community of fax users with whom you communicate.

If you use unattended communications techniques routinely with a lot of fax machines and it would be embarrassing or harmful for the wrong information to get into the wrong hands, you should explore the various levels of transmission security outlined in Chapter 2.

If you wish to send or receive quotes in a competitive bidding situation, you will want a level of security that at least ensures that the information won't be transmitted to the wrong machine. If you plan to send confidential internal information within a company, investing in the highest level of security may be prudent.

However, even if the information is sent to the correct machine, fax messages are visible to anybody standing near the machine at the moment. In essence, fax messages are like mail without an envelope.

That flaw can be eliminated by the use of a *secure mailbox*. When a fax is received, instead of automatically printing, it is stored in the machine's internal memory and is printed only when the correct password is entered. However, this feature generally works only between machines made by the same manufacturer.

The third security measure is *encryption*. Encrypting data prevents electronic eavesdropping as data move along the telephone lines. The message is scrambled when sent and a key or password is used to unscramble it at the receiving end. Encryption is required only for strict security. Unless you're in the CIA, you probably don't need it!

RS 232 Port

The RS 232 port can be either your connection to the future or a waste of money. In their current configuration, computers and faxes need not be linked, and unless you intend to integrate the two, an RS 232 port isn't necessary.

On the other hand, the new plain-paper, laser-printer fax machines create many opportunities. If you buy a laser-printer fax machine and have computers, an RS 232 port enables the fax printer to double as a computer printer.

Even with thermal-paper faxes, however, an RS 232 port might be a useful feature if you're a telex or TWX subscriber. The RS 232 port, along with a special modem and a service called Economy Telex, can drastically reduce your telex monthly bill. (For more information, contact Western Union or your fax representative.)

Many fax salespeople don't understand the purpose of the RS 232 port. This will change as computers and fax continue to combine features. (To find out more about computer/fax integration, see Chapter 5.)

Activity Reports

Activity reports are tools for managing fax usage. In small businesses or among personal fax users, activity reports are probably unnecessary. Telephone charges constitute the primary cost of sending fax, and those costs can be controlled through an analysis of the telephone bill.

However, activity reports are essential for monitoring actual fax flow. Activity reports indicate the time, date and duration of every fax message sent and received. The reports allow a manager to judge whether the fax machine is being used according to company guidelines. In some situations, for example, sending a large document via overnight mail makes more sense than tying up a phone line and someone's time while the fax is being transmitted.

Activity reports also can be important if people have free access to the fax machine. For example, through the activity report, one company discovered that one of its engineers was sending faxes to a competitor after hours. Not surprisingly, the engineer hadn't logged those transmissions in the written journal each fax user was expected to maintain.

MODE	CONNECTION TEL	CONNECTION ID	START TIME	USAGE T.	PAGES
RX	886 2 7357956	G-3	08/29 19:07	01'46	01(00)
TX	16037421944	G-3	08/30 07:45	00'45	01(00)
TX	0114414999767:::	G-3	08/30 07:56	00'58	01(00)
TX	16883909	CH PRINTING	08/30 11:19	01'02	01(00)
TX	16883909	CH PRINTING	08/30 12:37	00'49	01(00)
TX	12058776422	G-3	08/30 13:38	01'38	02(00)
RX	45 2 879460	G-3	08/31 06:58	01'20	01(00)
TX	1215561626B	G-3	08/31 09:45	00'59	01(00)
TX	17034338339	BANTA H-BURG UA	08/31 10:49	01'35	03(00)
TX	011452879460:::	G-3	08/31 10:56	00'47	01(01)
TX	011452879460:::	G-3	08/31 10:59	00'56	01(00)
RX	1818307577G	G-3	08/31 17:12	02'25	03(00)
RX	6814137	G-3	08/31 17:27	01'12	02(00)
RX	021 706 3301	G-3	09/01 07:31	01'27	02(00)
TX	1707684018A	G-3	09/01 07:36	04'57	08(00)
TX	0116126814137:::	G-3	09/01 08:00	00'56	01(00)
RX	021 706 3301	G-3	09/01 08:03	01'37	02(00)
RX		G-3	09/01 09:24	03'22	02(00)
TX	16883909	CH PRINTING	09/01 10:36	02'01	01(00)
TX	16037421944	G-3	09/01 10:41	03'56	10(00)

MAJOR PRICE GROUPINGS

People who read fax advertisements know that fax prices are dropping rapidly—so rapidly, in fact, that machines may not fit exactly in the price categories described below. However, fax features probably will continue to be clustered in much the same way, and prices will vary accordingly. With full-featured faxes dipping below the $3,000 mark and basic units costing less than $700, selecting the appropriate features is as important in choosing a fax machine as price.

No-Frills Fax

The common denominator among fax machines selling for less than $700 is that most are only Group 3 compatible and come with very few advanced features. Most print on 98-foot rolls of thermal paper and can receive transmissions unattended. Those two features are the very minimum you should demand in a fax machine.

Beyond this basic configuration, fax machines that cost less than $700 vary widely. Many do little more than send and receive. Most don't include document feeders. And although they may come with simple telephone handsets, most don't include auto-dialers. Because this group of machines often lacks a display screen, it's impossible to know whether you've called the right number. Not surprisingly, these machines have no activity-reporting features.

This class of machine is most suitable for light, personal fax usage and for cost-sensitive buyers. For example, if you periodically work at home and your office has fax capability, a low-end machine might be appropriate for those days when you commute only as far as your study.

However, price-conscious buyers should avoid specials that advertise "Brand Name, New Fax" for $399 or less. Often these machines are old Group 2 faxes, which can take three minutes to transmit a page—as much as 15 times longer than Group 3 faxes. Slow speed alone makes them a bad buy, not to mention their lack of compatibility. The fax revolution really began when Group 3 became available. Don't buy obsolete technology.

Also, be warned: Some Group 3 machines use slower 4,800-baud modems that perform at one-half the speed of most Group 3 machines. These machines can be best for very light use and if your fax budget is limited.

Middle-Range Fax

Most machines in the $700 to $1,200 range come with a liquid crystal display screen, transmit and remote terminal identification and the ability to generate basic activity reports. All communicate at the rate of 9,600 baud and are Group-3 and Group-2 compatible. Some are Group 1 compatible as well. Most have auto-dialing capabilities.

The most distinguishing features of machines in this category are automatic document feeders, automatic paper cutters, polling and internal memory. You should look for a machine with an automatic document feeder that can handle your needs. If you plan to send faxes with multiple pages, the automatic document feeder significantly reduces operator time. Although machines in this price range can poll other machines, they usually lack security features.

These middle-range fax machines can be tricky to buy. In the sales literature, they appear to have all the features of more expensive machines. But because some features are limited, others cannot be used effectively.

For example, many machines offer delayed transmission. But if the machine can be programmed to transmit only one document and you need to send more than one fax each night, you're severely limited.

Small automatic document feeders can have the same kind of impact. If you need to use the delayed-transmission feature to send a 20-page document, a 10-page feeder will make that feature useless.

Finally, in addition to the small document feeding capacity, many machines in this range have two serious drawbacks which limit their usefulness for high-volume applications. Most work only with 98-foot rolls of paper and lack automatic document cutting. If you receive a lot of faxes at an unattended machine, the lack of automatic document cutting may force you to spend more time cutting and collating messages than it took to transmit them. And unattended machines run out of paper often.

Mid-range faxes are most suitable for low- to medium-volume workplaces, such as branch offices, small companies and professional practices, traveling and trade shows, and personal fax applications. Without a large investment, they provide an excellent way to learn how fax can be used in your specific circumstances.

Full-Featured Fax

Full-featured fax machines should be able to serve as the central fax unit in any organization. In the $1,500 to $3,000 price range, you can expect to get every feature currently available. The most useful features are large paper rolls, automatic document cutting, automatic document feeders that hold 30 to 50 sheets, and internal memory.

In addition, many models in this range have such advanced features as secure polling and secure mailboxes, broadcasting and high-speed transmission. Of course, many of the most advanced features will work only with other machines made by the same manufacturer. If you need these features, be sure your machine and the ones you'll be sending to and receiving from were made by the same company.

High-end fax machines soon may come equipped with laser printing. Such models will be able to double as computer printers and will offer the advantage of copying on standard bond paper.

Deluxe Fax

This category will be dominated by equipment that combines several tasks, such as high-speed photocopying or laser printing with fax communication. These machines, which typically cost more than $3,000 per unit, have large internal memories and every feature discussed in this book. In the future, deluxe machines will be Group 4 compatible, which means they'll work on ISDN (digital transmission networks) and will be able to send an error-free page of information to another Group 4 machine in less than five seconds. When Group 4 machines first appear, they'll demonstrate the promise of the future rather than fill the needs of today—so buy with caution.

WHERE TO BUY

As the price of fax technology has dropped, the distribution channels for equipment have multiplied. Like copier machines, fax machines now can be bought directly from salespeople employed by the manufacturer, as well as from retail stores specializing in office equipment. Some high-volume consumer electronics stores, discount operations and even mail-order companies carry fax equipment. In response to increasing consumer demand, department stores also are stocking fax.

It's important, however, to buy your equipment from a channel of distribution appropriate to your needs. To a large degree, the appropriate channel is dictated by the price range you choose. Discount chains and mail-order companies usually offer the least expensive machines. Manufacturers' representatives offer the most expensive, because you're also buying that company's expertise and, often, its ongoing service and support. In some settings, this may be unnecessary. In others, it's vital.

Manufacturers' Sales Representatives

There are several advantages to buying from a manufacturer's sales representative. These salespeople are likely to know and understand their company's various fax configurations. They should be able to spend enough time with you to learn your business, so they can help determine which configuration best suits your needs and provide after-sales training to make sure you fully utilize all the features. They should be able to offer a wide range of leasing and financing options. (For more information on leasing, see Appendix A.) Finally, the sales force provides a direct link to the company's service organization. In short, a direct sales organization should be able to take care of you best.

But that service has its price. First, to explore a complete range of offerings, you should talk to representatives of several manufacturers, which can be extremely time-consuming. Second, sales personnel generally ask for the full suggested list price, whereas other channels offer substantial discounts. Finally, because salespeople usually work on commission, they want to sell the most expensive equipment.

Office Equipment Stores

Because office equipment stores generally carry a range of machines from several manufacturers, comparison shopping is easy. And because they're in a retail environment, they may offer competitive price discounts, particularly on older machines. In addition, many offer their own installation and training services.

Nevertheless, office equipment stores have their own drawbacks. First, you're not dealing directly with the manufacturer, and if problems occur, the retailer may not be able to solve them. Second, because the store's salespeople sell a wide range of products, they may not have in-depth product knowledge. Moreover, salespeople probably won't have time to assess the needs of your business. And they may steer you toward the units they want to sell rather than those most appropriate for your needs and budget.

Hence, if you choose to shop at an office equipment dealer, you must do more homework and understand what you want from your fax network. And you should be sure to check on warranties, service contracts and the location of the nearest repair shop, in case of a malfunction.

Discount and Mail-Order Companies

At discount and mail-order companies, the big benefit is low price. But you can't expect the salesperson to know anything about the features of a specific model or the differences between models. Be prepared to provide for your own setup, installation and training. However, for low-volume, personal fax users, the savings may compensate for the lack of follow-up and support, particularly for people with considerable electronic expertise.

FEATURES ASSESSMENT CHECKLIST

At this point, you should have a good idea of what you want in fax technology, how much you should pay for a machine and where to buy it. The checklist below can help you evaluate what best suits your needs. Be sure to take it with you when selecting fax equipment.

FEATURES ASSESSMENT CHECKLIST

COMPATIBILITY

	CRITICAL	OPTIONAL	UNNECESSARY
Group 3	❏		
Group 2	❏	❏	❏

Reasoning: _____Number of Group 2 receivers

RECEIVING CAPACITY

	CRITICAL	OPTIONAL	UNNECESSARY
Large	❏	❏	❏
Small	❏	❏	❏

Reasoning: _____ Average number of faxes received daily

_____ Average number of pages per fax

_____ Hours machine is unattended

_____ Paper capacity needed for unattended operation

AUTOMATIC PAPER CUTTER

	CRITICAL	OPTIONAL	UNNECESSARY
	❏	❏	❏

Reasoning: _____ Average number of faxes received daily.

_____ Average number of pages per fax

DOCUMENT FEEDER SIZE

	CRITICAL	OPTIONAL	UNNECESSARY
Large	❏	❏	❏
Small	❏	❏	❏
None	❏	❏	❏

Reasoning: Plan to use delayed transmission or polling? Yes_____ No____

_____ Number of pages of average fax message

TELEPHONE HANDSET

	CRITICAL	OPTIONAL	UNNECESSARY
	❏	❏	❏

Reasoning: Is a convenient alternative to integrated handset available? Yes____ No ____

DOCUMENT SENDING SIZE

	CRITICAL	OPTIONAL	UNNECESSARY
Large	❑	❑	❑
Standard	❑	❑	❑
Small	❑	❑	❑

Reasoning: _____ Maximum width of paper to be scanned

_____ Number of oversized documents to be transmitted per month

Is reduction copier accessible?
Yes____ No____

COPIER OPTION

CRITICAL	OPTIONAL	UNNECESSARY
❑	❑	❑

Reasoning: Do you want to preview documents before they're sent?   Yes _____ No _____

Would low-level photocopying capability be helpful?   Yes _____ No _____

RESOLUTION

	CRITICAL	OPTIONAL	UNNECESSARY
Fine Mode	❑	❑	❑
Super Fine Mode	❑	❑	❑

Reasoning: Do you need more than fine mode resolution?   Yes____ No _____

Will you communicate only with fax machines of the same brand?
Yes____ No____

HALFTONE TRANSMISSION

	CRITICAL	OPTIONAL	UNNECESSARY
	❏	❏	❏

Reasoning: Do you need to send photographs or grayscale graphics?   Yes_____ No_____

SUPERFAST MODE

	CRITICAL	OPTIONAL	UNNECESSARY
	❏	❏	❏

Reasoning: _____ Volume of international fax traffic in pages per month

_____ Number of fax receivers that support superfast modes

_____ Number of fax receivers from same manufacturer

_____ Speed of fax transmission in actual time test

DELAYED TRANSMISSION

	CRITICAL	OPTIONAL	UNNECESSARY
	❏	❏	❏

Reasoning: _____ Volume of faxes per month to be sent at low-cost hours

DIALING CAPABILITIES

	CRITICAL	OPTIONAL	UNNECESSARY
Auto-Dialing	❏	❏	❏
Speed Dialing	❏	❏	❏
Alternate Dialing	❏	❏	❏
Stored Capacity	❏	❏	❏

Reasoning: _____ Number of fax numbers accessed regularly

_____ Frequency of calling regular fax numbers daily or weekly

How busy are the numbers, particularly during unattended hours?_____

_____

_____ Number of telephone numbers that can be stored in memory

_____ Who will actually program the numbers?

STORE AND FORWARD

CRITICAL	OPTIONAL	UNNECESSARY
❏	❏	❏

Reasoning: Will the information you receive be passed on without comment? Yes_____No_____

BROADCAST CAPABILITY

	CRITICAL	OPTIONAL	UNNECESSARY
	❑	❑	❑

Reasoning: _____ Current broadcasting tasks

               _____ Potential broadcasting tasks

POLLING

	CRITICAL	OPTIONAL	UNNECESSARY
	❑	❑	❑

Reasoning: Will you be gathering information from many sources in limited time frames? Yes ____ No ____

How often? _____

Will you conduct two-way fax communication? Yes ____ No ____

AUTOMATIC VOICE/DATA SWITCH

	CRITICAL	OPTIONAL	UNNECESSARY
	❑	❑	❑

Reasoning: Anticipated monthly fax volume _____

Is a dedicated line cost-effective? Yes ____ No ____

Will you be present when faxes are received? Yes ____ No____

SECURITY AND SECURE MAILBOXES

	CRITICAL	OPTIONAL	UNNECESSARY
Secure Polling	❏	❏	❏
Secure Mailbox	❏	❏	❏
Encryption	❏	❏	❏

Reasoning: How sensitive and confidential are the data you plan to send and receive?_____

_____

Can you be sure that the intended person is the only one who will have access to the information?   Yes _____ No _____

How closely will you monitor actual fax sending and receiving? _____

_____

RS 232 PORT

CRITICAL	OPTIONAL	UNNECESSARY
❏	❏	❏

Reasoning: Will I use the fax for telex service?
Yes_____ No _____

Is this fax machine an effective printer?
Yes _____ No _____

What benefits will I receive today?_____

_____

_____

What future benefits will I realize from this machine's RS 232 capability?_____

_____

_____

ACTIVITY REPORTS

	CRITICAL	OPTIONAL	UNNECESSARY
Communication journal	❑	❑	❑
Transmission report	❑	❑	❑

Reasoning: _____ How many people will have access to the fax?

_____ How damaging can unauthorized fax usage be?

MOVING ON

By now you should have a good idea which features—and clusters of features—you need. You also should have an estimate of your price range.

The remaining chapters assume you own a fax machine. They provide information on installing, managing and maintaining this new technology.

CHAPTER 4

UP AND RUNNING

A FAX MACHINE FRESH OUT OF THE BOX

After you've decided to buy a particular fax machine, how difficult is it to get that machine up and running? In most cases, not too difficult. With a simple machine, you should be able to fax Singapore within an hour after taking the machine out of the box.

Unlike computers, which require days (if not weeks) of training, fax machines can be set up and put to work with a minimum of confusion, provided a few precautions are taken. On the other hand, if you don't work through installation problems carefully at the outset, you may deprive yourself of several advantages.

Key Steps to a Smooth Transition

Getting your equipment to work may be the easiest part of putting fax into operation. The difficult part lies in how you handle the following steps:

1. Decide where to place your fax machine and who will operate it.

2. Choose a long-distance telephone company.

3. Have your local telephone company install an additional telephone line.

4. Install your fax equipment and test it immediately.

5. Train your personnel, and encourage fax use by explaining its capabilities.

WHERE TO PUT IT? WHO WILL USE IT?

Your decision about where to install your fax should be determined by how it will fit into your communications network. How do you expect it to be used, and who will operate it? Do you want the fax machine in your own office, in a co-worker's office, in the mailroom, next to the copy machine, close to the receptionist or all by itself?

Location in Large Offices

In large offices, a single fax machine often is shared by 25 to 50 users. In such situations, it's usually located in the mailroom or in another central location with its own clerical staff. When a message is received, the mail clerk delivers it to the appropriate person.

Although this arrangement prevails, and has been used for telex communications for years, it probably won't be effective in the long run. Unlike telex, the fax machine's operation must be routinely monitored, because most fax machines don't ring bells to signal the arrival of a message. Also, the high cost (and low capacity) of telex limits its use. However, as fax use increases and prices drop, machines will be shared by fewer people receiving more messages.

In a large company, the mailroom still may be the most logical location for the fax machine. After all, fax closely resembles mail. But to handle fax effectively, the mailroom must be geared up and staffed properly. That may require extra personnel to deliver fax to the recipient promptly.

Moreover, a company may insist that fax messages be packaged in interoffice mail envelopes or special envelopes designating fax mail to help ensure confidentiality. Designing special envelopes for fax mail can have the added benefit of stimulating fax usage. (Some people theorize that Federal Express became a fixture in office communications in part because of its distinctive envelopes, heralding the urgency of the communications.)

Location in Smaller Companies

In small and mid-size companies, another logical location for the fax machine is near the central telephone switchboard operator, who often is responsible for sorting incoming mail. Both fax and telephone messages can be distributed the same way, and most employees are already in the habit of checking routinely for telephone messages. If fax messages are placed in distinctive envelopes, recipients passing by the switchboard will see they have an urgent message waiting for them. Thus, the distribution burden is shared by the recipients and the switchboard staff.

On the other hand, if you locate the fax machine near a switchboard operator who doubles as receptionist, be sure that the machine is placed where casual visitors can't access it. Remember, faxes don't arrive incognito. An inquisitive guest could read highly confidential information from a fax placed in the reception area.

If you plan to photocopy all incoming fax messages routinely, you may decide to locate your fax next to your copier. The potential advantages and drawbacks of the reception area location also apply to the copier area. First, like the reception area, the copier area is a high-traffic area. Second, the amount of time devoted to handling fax can be reduced by dovetailing it with other tasks.

On the other hand, unauthorized personnel can have access to incoming messages. Furthermore, unless you have personnel regularly stationed in the copy room, someone will have to check periodically for new fax messages—an arrangement that could prove clumsy and unreliable.

Rules of Thumb for Fax Machine Location

Because fax machines operate almost noiselessly, they can be placed almost anywhere without interfering much with normal activities. Unlike computer printers, fax machines don't have to be encased in a plastic box to reduce noise. Consequently, the rule of thumb in finding a location for a machine can be its convenience for the people who plan to use it. If the fax machine is poorly located, people probably won't use it as much as they would if it were placed in a convenient spot.

Because larger businesses have more sophisticated in-house mail systems, it may be beneficial to integrate general fax traffic with procedures for distributing mail.

A mid-range solution is to deploy fax at a departmental or work-group level. One of the incoming telephone lines to that department can be reserved for fax use, and the departmental clerical staff can be responsible for distributing fax messages. Without eliminating the problem of confidentiality, strategic distribution of fax machines throughout an organization often makes more sense than centralization.

Your decision, of course, has important ramifications. The more central the fax location, the more you must rely on clerical staff to operate the machinery and distribute the messages. That means staff must be trained and sometimes redeployed from other responsibilities. The biggest risk, however, is that if the clerical staff doesn't perform well, fax won't be used well and you won't enjoy its full advantages.

Personalizing fax has drawbacks as well. First, higher-paid employees will waste time fiddling with the machine. Moreover, the greater the number of users, the more messages each user must send to justify the placement, which can result in unnecessary use.

Last, if a personal machine is perceived as a status symbol, jealousy and rivalry can develop. Good fax equipment isn't inexpensive. When deciding the location of the fax machine, decide how and by whom you want the machine to be used.

CHOOSING LONG-DISTANCE SERVICE

Choosing a long-distance service for fax use can be as difficult and confusing as choosing a long-distance service for regular telephone use. Of the long-distance providers, none yet offers special inducements for fax users.

The first and most obvious question is whether the carrier serves the locations to which you'll be sending faxes. Even some of the large long-distance suppliers still have limited international coverage.

Determining the quality of transmissions sent by alternative long-distance suppliers is also important. The human ear adapts to and tolerates the quality difference between fiber optic and microwave transmissions. Fax is more sensitive, and poor quality transmission forces it to fall back to a slower transmission rate.

For example, if lower quality long-distance service is 20 percent less expensive per minute, but your faxes must be sent at 4,800 baud instead of 9,600 baud because of the lower quality transmission, you will wind up spending 60 percent more in telephone costs, instead of saving 20 percent!

Furthermore, fax messages may be garbled and hard to read.

The final element in the selection of a long-distance service is availability of special services. Bear in mind that WATS lines can be used as effectively with fax technology as with regular telephones.

After the AT&T divestiture and deregulation of 1984, most businesses re-evaluated their long-distance suppliers. If you're satisfied with your present supplier, you probably should add the fax equipment to your current network. Otherwise, purchasing fax equipment can provide you with a good opportunity to reassess your long-distance options.

After you've decided where you want to place fax equipment and which long-distance service you plan to use, you may need to install additional telephone lines. Because new installation often requires several weeks' notice, call the phone company promptly and keep the following points in mind:

1. If you can justify the cost, reserve a telephone line dedicated to fax use. Although a fax machine can accept calls through a PBX system, separate lines allow fax machines to answer calls unattended. They also prevent scrambled messages or other difficulties that result when people unwittingly pick up the phone while the fax is in use.

2. You don't need a special line—such as a data line. Fax works just fine over regular telephone lines. Be suspicious if a salesperson offers a special line for fax use.

3. If you're installing a fax in your home office, ask for a residential line, not a business line. A residential line is usually less expensive.

4. If possible, don't have your fax number published in the phone book. If you do, some people may call directory assistance for your phone number and instead receive your fax number. That can be ear-splitting, as well as frustrating.

5. Think twice about having your fax number listed in any kind of directory. You may be deluged with junk mail.

6. Don't get additional features on the telephone line you plan to use for fax. Many local Bell operating companies automatically include a package of features for the subscriber to "test" with each new line. Be sure to insist that no features are added, particularly Call Waiting, Automatic Call Forwarding and so forth. Above all, avoid Call Waiting, which signals another incoming call with a loud click on the line. The fax machine doesn't understand the function of this click and misinterprets the information, consequently garbling incoming and outgoing messages.

Fax technology's sensitivity to Call Waiting may present problems for people who want to install fax on an existing phone line also used for voice communication. If you can't have a dedicated fax line, the best solution is to install a second line with a rotor. Incoming calls then are routed from a busy line to the second line. Rotor capability costs only a few dollars per month—not significantly more than Call Waiting—and it's more convenient and not threatening to your fax.

7. Determine the best call-area billing system for your fax phone line. If most of your messages will be local, you may want to invest in extended local-area dialing. If most of your faxes will be sent long-distance, you probably will want a restricted billing area. Before calling the telephone company, analyze your needs carefully to decide which of its billing options is appropriate for you.

INSTALLING YOUR EQUIPMENT

The procedures for installing and using your fax machine depend upon where you bought it, the complexity of your system and the number of people who will be using the system.

The Manufacturer's Representative

If you bought your machine from a manufacturer's sales representative, installation and training should present no problem. In the purchase agreement, you should insist that the representative come to your workplace, install the machine and train your personnel. Because you've paid a higher margin for service, don't waive that advantage, particularly if you've purchased a complex machine with such features as broadcasting and account codes and if you expect many people to have access to the fax.

As a rule, try to have a competent representative explain and demonstrate as much as possible about the machine you buy. Unfortunately, because operating procedures vary markedly from manufacturer to manufacturer, you can't count on other fax users to help resolve problems that may arise.

Not surprisingly, some sales representatives are more interested in making sales than in providing support. So you must determine exactly how the support you've purchased will be delivered. As people who buy new cars ruefully learn, the salesperson who says he or she will be your personal representative to the company is too often unavailable.

Nevertheless, company support and service personnel should be able to install your machine and train your employees better than anyone else. If you make a large unit purchase, try to have an account support representative assigned to you. If you buy direct from the manufacturer (and the extra level of available support is a good reason to go in that direction), you may want to choose a vendor with support activities separate from sales activities.

Office Equipment Dealers

If you buy your fax machine from a retail store, try to elicit as much information as possible about the machine *before* you buy it. When you leave the store, you may be on your own. On the other hand, many retailers pride themselves on the product knowledge of their sales personnel. Try to test that knowledge and put it to use.

Before completing the sale, open the box at the store and inspect the machine. Then browse through the instruction manual. The quality of these manuals varies widely. Some are virtually unintelligible. Others provide users with clear and concise guidance.

Regardless of the quality of the instruction manual, ask the store personnel whether they can program the transmit-terminal identification number and the time and date into the memory. If no one at the store can perform these tasks, you may want to shop elsewhere.

In addition to one-time operations (such as programming the transmit-terminal identification number), ask a salesperson to review the basic operation as you follow along with the manual. Have him or her program some numbers into the speed-dialing memory and watch as you program some. Review the paper-loading procedure. Don't leave the store without a roll of paper loaded into your machine.

Mail Order or Discount Stores

If you bought your machine from a mail order company or a discount store, you decided to invest your own time in exchange for a low price. With fax technology, that can be a good investment. When the machine arrives at your workplace, unpack the unit carefully and inspect it for damage. Then review the instruction manual, plug the machine in and go through the initial programming procedures for transmit-terminal identification, date, time, speed-dialing and so forth.

Reserve enough time so that when you finish, your machine will be fully operable. Because you basically can plug a fax machine into a phone jack and send a message right away, you should program your features immediately; otherwise, you may never program them. If you've paid for advanced features, it's a waste of money not to use them.

AVOIDING POWER OUTAGE PROBLEMS

Power outage problems pose one of the most serious threats to your fax machine. A power surge or lightning can disable your fax and result in a sizable repair bill. To protect your fax from being damaged, use a power surge protector. A good one costs under $50 and provides excellent insurance against power problems.

Speaking of power protection, an Uninterruptable Power Supply (UPS) is a great power protector and lets your fax keep operating even if the power fails (although you'll need a flashlight to read incoming faxes). This option need only be considered if you regularly experience power outages and if it's important that you always be able to receive and send faxes.

THE BUCK STOPS HERE

No matter where you buy your equipment, assign responsibility for its operation to one person who'll maintain the equipment, train operators and generally stay abreast of the field. Fax technology is improving rapidly. One person in your organization should be responsible for keeping you on the leading edge of communications.

ADDITIONAL RESOURCES

If you're heavily invested in fax equipment, you may want to keep up-to-date on new product developments. *SpecCheck Fax* from Dataquest (800-624-3282) gives manufacturers' specifications on several hundred fax machines and is updated regularly. You can subscribe to it or order single copies. Datapro (800-328-2776) offers manufacturers' specifications coupled with user surveys; its publications are oriented to large businesses. Finally, *What to Buy for Business* (914-921-0085) publishes a fax guide and periodic updates.

THE FINAL STEPS

After you've installed your fax machine, test everything. Because you may be relying on several different vendors—one for fax hardware, one for telephone-line hardware and one for long-distance services—each probably will point the finger at the other if something doesn't work. All glitches should be resolved before you print your fax number on your business card. Lost faxes can hurt your business even more than lost telephone calls.

Finally, take time to train your personnel—all your personnel—on fax usage. To get the most out of fax, not only must the equipment be running but your employees must be running it. Training should consist of a one-time introduction to the new equipment, followed by ongoing suggestions about how to use fax and when not to use it.

MOVING ON

Now that you've installed your fax equipment, all you have to do is sit back and wait for the cost-savings to come pouring in, right? Wrong! In the next chapter, you'll discover how to best manage fax within your company and stay informed about its many changes.

CHAPTER 5

MANAGING YOUR FAX

In the early 1980s, midlevel employees in many corporations began to discover the potential of personal computing. If management was slow to react, these corporate renegades used personal computers to do their own jobs faster and better, often purchasing machines on their own and using them for a single application, such as word processing or spreadsheet analysis.

Within a few years, many corporations were (and still are) littered with incompatible machines, running dozens of different software packages for the same applications. The power of personal computing ran uncontrolled. Training, service and software costs soared until companies decided to standardize on a single machine. Even if each employee used a computer effectively, personal computing often wasn't used effectively throughout the company.

Now fax technology has reached a similar stage of development. Individuals and departments are realizing independently the advantages of distributing hard copy instantly. In many companies, the accounting department may want a fax machine for sending reports to and from company headquarters. Then the sales department may begin to use the machine, but worry that accounting is unduly privy to messages and thus want a machine of its own. In large sales departments, a single machine may not suffice.

High-volume salespeople may want their own machines.

Without central coordination, fax technology can wend its way through a corporation without much thought, control or care given to its overall use. To maximize its technological advantages, fax must be well used and well managed.

Because every business is different, there isn't a formula for introducing and managing fax use. Nevertheless, in most situations, even in home offices, resolving critical management issues will make fax operation easier on everyone.

Fax technology raises three basic management issues: *information* management, *personnel* management and *cost* management. Information management includes procedures and processes that ensure that fax enhances the flow of information within your company. Personnel management enables the right people to use and maintain fax and the fax network. Cost management ensures that the new technology saves rather than squanders money.

Let's examine these three areas of fax management.

INFORMATION MANAGEMENT

Information management can be divided into four areas: integrating yourself into the fax community; ensuring that the information you *send* reaches its intended destination; ensuring that the information you *receive* reaches its intended destination and is properly stored; and, finally, establishing guidelines defining the appropriate uses of the fax machine.

Entering the Fax Community

Once your fax is up and running, your first task is to inform those you want to communicate with that you have a fax machine. This seems obvious, but if you mishandle the way you inform them, you may hinder the development of fax as a routine communications instrument in your setting.

First, decide who should know you have fax. If you bought the machine for your sales department, do you want your vendors to know you have it? If you bought only one or two machines for your entire operation, are you sure that everybody with whom you do business needs to have instantaneous access to your company via fax? At times, it may be advantageous to have some people send you information by mail rather than by fax.

For example, many public relations agencies now fax their press releases to news agencies and their clients. They believe that fax gives the information an aura of urgency and importance—and it may do just that. However, journalists usually work on deadline and cannot react quickly to random press releases. Instead, they generally scan their mail periodically for new ideas. Because they can't be easily integrated into the journalist's work routine, press releases received by fax can be a nuisance.

On the other hand, journalists often need background information on short notice from companies or institutions. For example, a large daily newspaper in California has bureaus more than 100 miles away from the main editorial location. When a reporter requested clippings from the newspaper's library, those clippings used to be sent by courier service, which was slow and inefficient. Now they're sent by fax.

Thus, although journalists may want information sent via fax in a specific situation, they may not want every public relations agency to know they have fax capability.

Getting the Word Out

Before you let the whole world know you have fax, decide if you want the whole world to know. Of course, once you've determined the community of people with whom you want to communicate via fax, you have to let them know your fax number. The easiest, most efficient, comprehensive way to inform them is to send an announcement via the fax machine. The announcement can be short and to the point:

"ABC Company is pleased to announce that its new facsimile number is 888-444-4444. Please add this listing to your internal fax directory and route this information to the proper departments. Our regular mailing address is 5 Main Street, Centertown, NC 58888, and our voice telephone number is 888-444-8888." The inclusion of your mailing address and telephone number allows recipients to photocopy all the information and update their telephone directories.

Of course, a simple message via fax probably won't be sufficient to inform everybody you have fax. Add your fax number to the business cards and stationery of the people who are responsible for authorizing fax usage.

Once again, the fax number should be restricted to those who will use the fax network. By restricting fax use to its intended purposes, you can control the growth of fax within your organization and force employees to justify their demand to be included in the network.

Increasing Your Fax Audience

Once you've identified your fax communication audience, try to draw them into your network as many ways as possible. In fax technology, it's better to attract too many people than too few. Because fax technology is so effective, it quickly can become overused and overburdened. And although you want to restrict fax usage to appropriate purposes, controlling it too severely can prevent you from receiving the maximum return on your investment. If usage gets too heavy, more machines can be added to your network. But if fax usage is too light, you can never measure the business you may be missing.

For example, a fax machine can be a quick, reliable 24-hour-a-day order taker. A customer can complete the order form and fax it to you in a fraction of the time it takes to order by phone or mail. Placing your fax number in large print on your order form encourages fax ordering.

To make it even more convenient, and if your sales volume can support it, offer an 800 fax number. It's free to your customers, and they'll appreciate the speed and professionalism an 800 number implies.

Checking Your Transmissions

A fax message is like mail without an envelope. You may have no idea how a fax message is handled at the receiving end. Often it passes through several hands before it reaches the person for whom it was intended.

Always include a complete transmittal sheet with every fax message. Because the transmittal sheet is the most important tool to help get your message to its recipient, no fax message should be sent without one. The transmittal sheet allows anyone who receives a fax to make sure the entire message has arrived safely, collate it and forward it to the proper destination. The standardized transmittal sheet provides the following information:

1. Identity of the sender.

2. Identity of the recipient.

3. Length of the message.

4. Person to be contacted if there has been a transmission problem.

5. Additional instructions to the receiving operator, such as alternate recipients, cc:'s, and so forth.

The first element in the transmittal sheet should include your company name, address, telephone number and fax number. Then the receiving party can log the receipt of the information. Many companies use their letterhead for the transmittal sheet. If you choose to do that, it's prudent to include the company identification a second time in larger type, because faxed versions of stationery are sometimes hard to read.

The second element of the transmittal sheet often resembles a memo. It should provide space both for your name as sender and the recipient's name. Leave several spaces for the recipient's name if you want copies of the fax sent to several people. The fax operator who receives it can copy and circulate the information.

FAX

Date: _____

TECO U.S.
Chapel Hill, NC 27515-2496
Telephone (939) 732-9449
Telefax 939 142-3898

Attention: _____

Company: _____

From: _____

Number of pages_____including this page

Comments:
_____
_____
_____
_____
_____
_____

Next, you should note the number of pages, including the transmittal sheet, that have been sent. For many reasons—usually telephone noise—one page in a long fax document may be lost in transmission. Moreover, the communication might be garbled or otherwise rendered unreadable. So, this information lets the receiver make sure that the complete document has arrived.

Don't forget that the page numbers in fax transmissions don't necessarily conform to the page numbers of the document you're sending. For example, you may be sending pages 10 to 25 of a report, so always number the actual pages of the fax transmission. Furthermore, if you send your fax in sequence—that is, from page 1 to 9—page 9 will be the top page in the pile at the receiving end, and the transmittal sheet will be at the bottom. By including the page numbers on the transmittal sheet, you let the receiver know immediately if there's a problem. Many fax machines automatically number pages, as well as stamp them with the time and date, saving you this task.

Because fax information generally is urgent—or at least timely—you may want to list several alternative recipients for the receiving operator to notify. If only one person is listed and that person is unavailable, the delay may cause your information to lose its timeliness and effectiveness.

The bottom area should be large enough to include short notes and information. In some cases, the whole fax message may fit into the comments area. In others, the area can be reserved for special handling instructions, such as "Notify Jim at the Dallas office when this arrives," or "Immediate response requested."

Also, if you can't have your machine on 24-hour alert for incoming messages, be sure to specify on your transmittal sheet the hours when you can receive faxes.

A transmittal sheet doesn't have to be full-sized. A short cover sheet can convey the same information and require shorter transmission time. The 3M Company sells Post-It notes printed as small cover sheets. A custom version for your company could be very effective, particularly if you often send single-page faxes to the same people. However, the Post-Its will cover a few inches of the document. Another option is Auto Cover Sheet, which programs your machine to produce an automatic fax cover sheet, sent at the end of the transmission. It includes the number of pages and the sender's name. The recipient's name is also included, if your machine has the name stored with a number in dial memory. Remember, however, that entering names without a keyboard is slow and tedious. For this feature to be used effectively, you'll need a separate entry for each person at the company you're faxing to. If you don't use cover sheets now, this might be good to consider.

Backing up Your Fax

At most companies, mail-handling routines are more firmly established than fax-handling routines. Because you often can't be sure the recipient has received your fax, you can use one of several back-up methods.

For example, you can routinely mail a copy of a fax message after you've faxed it. Although it will arrive several days later, the letter often will be welcome. If your recipient has been out of the office for several days, the fax message may be lost or misplaced, but the mail will get through. Also, mail is easier to file than fax transmissions, in which cutting lengths are inconsistent; and thermal paper is more difficult to handle than bond paper.

A second strategy is to notify the intended receiver by telephone that a fax has been, or will be, sent. Though it may seem redundant to call and then fax a message, you'll often save time by reducing errors. In one company, for example, an executive found himself talking to an unusually demanding customer for hours on the telephone, because pricing information was misunderstood. It turned out the fax clarifying the information hadn't been received by the customer. Now the customer calls to alert the executive that he'll be sending a request for price quotes via fax. The executive finds the answers, calls and then faxes them back.

Similarly, a manager may call a vendor to request certain product information. Rather than sit at the phone waiting for the salesperson to fumble through data sheets, the manager can instruct the salesperson to fax the information and call as soon as he receives it. Instead of being redundant, fax and telephones often work best in tandem.

The Right Tool for the Right Job

Because fax transmissions often are seen by more than one person, you must be extremely cautious about the information you send. Some information can be damaging or have unintended consequences.

For example, one small company lost a bill from a supplier—an honest mistake. In this case, however, the supplier didn't send an overdue notice. Instead, an overeager accounts receivable clerk faxed a message threatening to cut off service if the bill wasn't paid in five days. A clerk at the receiving end interpreted the message to mean the company didn't have enough money to meet its obligations, including the payroll. Morale among the employees was shaken until the owner resolved the problem. But she was so angry at the indiscreet and aggressive way the overdue bill was handled that she immediately began to search for a different vendor for that service. In the wrong hands, even innocuous information can seem ominous.

If the information sent is indeed ominous, a faxed message can precipitate dramatic and undesired response. For example, a Los Angeles bank executive notified branch managers that the commission structure for the loan representatives would be reduced in the wake of the bank's poor quarterly performance. His intention was to have the managers inform the loan officers. However, for the sake of speed and efficiency, he sent the notice via fax. When they prematurely learned about the changes, more than 60 loan representatives quit, decimating the sales staff. This kind of information would have been handled best in face-to-face meetings or by telephone.

Protecting Confidential Information

Confidential information must be safeguarded. The fact that you have a top-of-the-line fax machine in a central location doesn't mean you have to channel all fax traffic through it. If you wish to fax personnel records and payroll data, for example, you may need a small fax machine located in a secure area. An alternative to buying a second machine is having one with extensive internal memory and secure mailbox features (see Chapter 2).

Checking Your Incoming Messages

Check for fax messages regularly. For fax to work well, a message must be received promptly, acted upon and, when appropriate, filed.

In most situations, fax messages will be received by a fax operator instead of the ultimate recipient.

Operators should be instructed to check the machine on a regular schedule.

The advantage of fax is its speed. If the machine is checked for messages only once or twice a day, that advantage is lost. Moreover, if incoming messages pile up, they'll be harder to sort and collate.

Because the technology is new, take time to establish routines to ensure that fax mail isn't lost either before or after it gets to the receiver.

Routing and Storing Fax Messages

The correct procedures for routing and storing incoming fax traffic depend both on your fax operator and your policy for handling other forms of incoming information, particularly mail. For example, many executives instruct their secretaries to copy every significant incoming letter and store the copy in a correspondence log as a safeguard if the original is lost or misfiled.

If you use that system, follow it for fax messages as well. If you have other methods in place to safeguard information, you'll want to integrate fax messages into those systems. If you don't have a system, you should establish one. And keep in mind that fax messages on thermal paper simply aren't as sturdy and easy to handle as those on regular bond paper.

Protecting Your Fax Documents

In most cases, when a message is received, the operator collates it, attaches the incoming transmittal sheet and enters it in a log. On the basis of their experience with telex, some companies copy every fax message that arrives and store the copy in a special fax file. Consequently, if a fax is lost, the information can be retrieved by using the transmission log to determine when it was received and pulling the message from the file.

Some managers object to this system because the file precludes confidentiality. In some ways they're right. Central copying and filing may infringe on privacy. A solution could be to automatically copy the fax when it arrives and deliver that copy with the original to the recipient.

If you choose not to adopt a policy of having the fax operator copy messages as they're received, you should instruct the designated recipient of a message to copy it immediately, especially if the information is to be kept for any length of time. This is particularly important for fax messages printed on thermal paper, which isn't as sturdy as bond paper, doesn't last as long and darkens when exposed to heat. As bond-paper fax machines come down in price, this problem will eventually disappear.

Streamlining Fax Distribution

Once you've established procedures for removing faxes from your machine and safeguarding the information you've received, you need to be sure that faxes are delivered promptly or picked up regularly by the intended recipient.

Senders expect their fax messages to get through quickly and should be told if those expectations can't be met.

There are several appropriate methods for ensuring prompt delivery. Fax messages can be delivered by the fax operator (in which case, special fax envelopes should be used); you can install fax mailboxes; messages can be distributed in telephone message slots (not an option if you use a voice-messaging system); you can buy a fax machine for every desk that receives fax messages, an expensive alternative.

Obviously, the option you choose reflects the way you do business and the fax network you've established. No matter how you distribute fax messages, you should establish clear guidelines and policies on responding to incoming fax messages if the intended recipient isn't in the office and for alerting people on the road when they've received a fax. Senders know that they've successfully sent a fax transmission, but can only assume that it has reached the intended recipient.

Devise ways to let the sender know if the recipient hasn't received the message. The most direct way is to send a return fax stating that the recipient isn't in the office and won't be available for a specified length of time. Of course, this requires that either recipients inform the fax operator of their schedule or their support staff handle the communication.

Fax Junk Mail

Not long after a small manufacturing company bought its first fax machine, the firm received an intriguing message via fax. A company was offered a $20 ice chest in exchange for providing the sender with the fax numbers of 40 companies. The sender stated that it wanted to fax promotional material about low-cost fax paper to those 40 fax users.

This incident illustrates a major trend and nuisance for fax users—fax junk mail. Because they work over telephone lines, fax machines are susceptible to the solicitations of charities and junk calls. Once marketers have your fax number, you can do little to prevent their sending information, particularly if you don't have security features. After all, your fax machine must be available for transmissions for regular business purposes.

One sure-fire way to receive a load of fax junk mail is to list your number in a fax directory. The logic behind fax directories seems as compelling as the logic behind telephone directories. If somebody you don't know wants to reach you via fax, a directory offers the easiest way to find your number. Or, as one directory advertises, "With a fax directory, your fax capabilities have no limit."

But, unlike telephones, fax machines aren't general-purpose communication instruments. For now, fax enjoys a degree of urgency. It often demands more attention than even overnight mail, and you don't want to have to sort through a lot of junk fax to find your message. Indeed, if your machine is jammed with unwanted fax messages, important messages can easily be lost.

Safeguard your fax number. Don't expose it thoughtlessly. Before you list your number in a fax directory, consider the consequences.

Just as you protect your own fax number, protect the fax numbers of others. By responding to a junk fax solicitation requesting a copy of your communication register, you're giving away dozens of other people's fax numbers.

If you want to stop fax solicitations, call the company sending them and ask to be removed from its list. You may want to prepare a form requesting that your number be taken off a sales list. You then can fax it to the soliciting party. Reasonable companies will comply with your requests. After all, they want your business, not your wrath.

Junk fax represents one of the greatest threats to the effectiveness of fax communication. If you take the precautions above, fax should remain a fast, inexpensive and effective method of communication.

Some states have begun to take a hard line against junk fax. Maryland and Connecticut, for example, have passed laws forbidding companies from sending unsolicited fax. This could signal a trend.

Nonetheless, you should carefully monitor who gets your fax number. Even in Maryland and Connecticut, a person who has your permission to send a fax once is no longer constrained by the junk fax regulations.

If junk fax does get out of control, new generations of fax equipment will probably be capable of screening incoming fax communications. One possible method might be to require pre-registration of all fax machines. Machines would have an allowed access list programmed with the fax numbers of all machines authorized to send messages to it. If a machine not listed tried to send to such a machine, it would be refused.

With appropriate software on PC/FAX, another method might be to let a one-page request for registration be transmitted from any machine. If the receiving machine operator didn't then program the machine's identification information into the access list, no further faxes would be accepted.

Internal Fax Misuse and Abuse

Outside abuse in the form of junk mail is hard to counteract. Inside abuse in the form of impractical or unnecessary use of the fax machine can be curbed by guidelines specifying the type of information that can be sent by fax—guidelines similar to those used for overnight mail. The rule of thumb is that messages should be faxed when they're urgent, when the quality of the paper they arrive on isn't critical and when they're short enough to be easily transmitted by the fax machine.

Because fax transmission is generally inexpensive, the guidelines need not be as stringent as those for overnight mail. Nevertheless, if a machine is to be used by the sales department to receive inquiries and send quotations, you don't want its time dominated by the central administration office checking supplies at branch offices. If that happens, you probably should buy another fax machine for your administrative staff.

Also, many people make the mistake of sending a 30-page report by fax, when it won't even be looked at until the following afternoon. In these situations, overnight letters can be less expensive and the recipient gets better-quality paper.

It's also important to set a policy about fax abuse. Be sure the staff understands that the machine is for authorized business use, not to fax requests to the local radio station or send cartoons to a friend in another city.

Use the fax machine to communicate only specific types of information. After you've bought a fax machine, the uses for it are likely to proliferate. Take time to develop procedures to ensure that you're integrated into your fax community and that the information you send and receive reaches its destination and is properly stored. Most important, be sure fax is restricted to its intended purposes, even as its usefulness to others grows.

Legal Considerations

One of the most commonly asked questions concerning fax is, "Is it legally binding?" That is, is a signed document transmitted via fax a legally binding agreement? The simple answer is, yes. But like any legal question, the answer is not quite so simple.

There are two basic requirements. The agreement must be in writing, and it must be delivered. Agreements telexed to another party have been found to be valid; but, unlike telexes, faxes include a signature.

While the agreement may be legal and binding, it's best to have the original signed document sent by mail. If you ever have to go to court, the "rules of evidence" require the best available evidence, which in this case would be the original document. You can subpoena the original, and if it can't be produced, the fax then can be used as evidence. While no specific challenges to faxes have been brought to a court at this time, xerographic copies of agreements have been upheld repeatedly when the original isn't available.

PERSONNEL MANAGEMENT

Although most fax machines can work unattended, you must attend to people using the fax. The way people use the machine determines its efficiency. Personnel issues usually involve two groups of people—those who use the machines and those who manage the fax network and fax usage.

Do You Need an Operator?

Fax transmissions are like weddings. You spend far more time preparing for them—preparing documents and transmittal sheets, looking up the appropriate fax number—and more time cleaning up afterwards—receiving the fax, copying and distributing the message, filing the information—than you do attending them—sending information over the telephone line. Someone has to take all those steps; the question is, who?

In general, fax operation is a clerical task. Consequently, the responsibilities for sending and receiving fax should fall to the clerical staff, either individual secretaries and assistants or a centralized staff. A centralized staff is perhaps more efficient for receiving fax messages. One person can monitor the machine or machines routinely for incoming messages, copy those messages and then forward them to the recipient.

Incoming fax is like mail, particularly in companies with a limited number of machines. One person or a small staff can handle all the incoming material and distribute it.

Outgoing fax, however, is like the telephone. People want to send their messages at a moment's notice. They don't want to be dependent on someone else's work routine or have their priorities reordered by someone else. Consequently, sending fax probably should be a distributed task, and several people in the organization should be trained to send fax. For outgoing messages, you probably won't want the fax machine to become the "turf" of one individual who's the only person trained to use the machine.

Although fax transmission is primarily a clerical task, all personnel who create fax messages should learn how to send them as well. And while you don't want managers and executives to waste their time operating the fax machine during the day when support staff is available, you may want them to be able to send faxes if they sometimes work at night or on weekends.

A two-step system seems best for fax transmission—one that reflects the dual nature of fax. One person or a select group of people should be responsible for monitoring the inflow of fax messages and logging and safeguarding the information.

Sending responsibilities, however, should percolate throughout the organization. The one exception to this system is delayed transmission. Because most fax machines have fairly limited delayed-transmission capabilities, one person should be in charge of queuing up the messages to be sent each night. To facilitate the control of delayed transmission, you probably will want to set up a file near the fax machine for messages scheduled for delayed transmission.

Company Fax Directories

One further issue concerns the establishment of a centralized internal fax directory, listing every number to which a fax is sent. On the one hand, such a directory institutionalizes your fax community and prevents the loss of fax numbers due to personnel changes. On the other hand, many people consider fax numbers confidential the same way they do personal telephone directories. They want to control access to the numbers they've accumulated. The only way to resolve this problem is to meet with the people who use the fax machine and make a decision. An internal directory, even if incomplete, is essential.

Encouraging Fax Use

After the fax machine has been successfully installed, the fax manager should establish a training schedule for all appropriate personnel.

When people start to use fax, the fax manager should be on the lookout for innovative uses of the technology and should publicize them in the company newsletter or by word of mouth. This publicity serves three purposes: it reminds people who're still doing business the old way that they should consider using fax; it allows fax to grow as part of your communications network; and, the more innovative you are with fax, the better the return on your investment.

Keeping the Machines Fit

The fax manager must be sure the fax machine remains up and running, which requires careful evaluation of the need for a service contract. With few moving parts, fax machines are generally reliable pieces of equipment; but some models, such as laser-printer-based machines, require periodic service.

Staying Up-to-Date

The core elements of fax technology probably won't change dramatically over the next several years. Nevertheless, as demand for fax grows and it becomes more of a mass-market item, you'll be able to get more features and lower prices. You need a fax manager to monitor continuously the possibilities and opportunities opened up by new generations of fax machines.

Some key developments you can expect in the next few years include an increase in memory, which will allow greater use of delayed transmission; widespread use of bond paper, which will provide better quality faxes than thermal paper, and may eliminate the need to copy every fax; and further integration of computers with fax machines, which will give faxes more intelligence and flexibility.

One person should be responsible for analyzing those changes and deciding when to upgrade the fax network. Because fax can pay for itself quickly, your company may benefit from upgrading regularly so that you always have the most appropriate, cost-effective technology.

COST MANAGEMENT

The fax manager's final responsibility should be cost control. Fortunately, calculating fax costs is a fairly straightforward procedure. The main tool is the activity log and communication journal, which details where each fax was sent and how long it took to send. The log can be collated with the telephone bill to determine the telephone charges associated with fax.

The other costs are supplies, equipment and operator time. In small operations, it's often most convenient to include fax costs within the larger categories where they fall—telephone costs, office supplies and personnel.

When your fax is in full operation, the amount of overnight mail you send should dramatically decrease. If your overnight mail budget remains the same or increases, your company either isn't using fax efficiently or is growing rapidly.

SOME RULES FOR MANAGING FAX: A SUMMARY

1. Before you let the whole world know you have fax, decide whether you want the whole world to know.

2. Restricting fax use forces employees to justify their demand to be included in the network.

3. When you've identified your audience for fax communication, try to draw them into your network through as many means as possible.

4. Use the fax machine to communicate only time-sensitive information.

5. Provide a complete transmittal sheet with every transmission.

6. Check for fax messages regularly.

7. Senders expect their fax messages to get through quickly. If those expectations can't be met, let them know.

8. If your message is confidential, act accordingly.

9. Use back-up forms of communication whenever appropriate.

10. If you plan to keep fax messages on file, copy them onto better paper.

11. Safeguard your fax number against fax junk mail. Don't expose it thoughtlessly or list it in a fax directory without considering the consequences.

12. Always sign your fax messages. As with letters, a signature makes the fax more personal and compelling.

13. If you can't have your fax machine on 24-hour alert for incoming messages, be sure to specify on your transmittal sheet the hours when you can receive fax.

14. Request fax numbers from those with whom you do business.

15. Don't demand instant response from the receiver unless you've agreed to a quick turnaround beforehand.

16. Keep fax numbers accessible to all employees.

17. Experiment. The payoffs can be dramatic.

CHAPTER 6

PC/FAX: FRIEND OR FOE?

Integrating fax with personal computing is the wave of the future in communications. However, while current technology combines the two, it's important to weigh the advantages and disadvantages before you add fax capability to personal computers.

If you don't have at least a rudimentary understanding of computer jargon; if such terms as megabytes, incompatibility, ASCII, half-slot boards and application programs mean nothing to you; or if *.pic and *.tif look like typographical errors, this chapter may make little sense. To recognize the potential and pitfalls of PC/FAX, you must have a basic understanding of personal computing.

COMBINING FAX WITH PERSONAL COMPUTERS

Over the next ten years, merging data processing and telecommunications into a completely integrated system will be a major trend in office automation. There's no doubt that the computer is the foundation of the automated office. People will not only continue to generate more and more documents via PCs, they'll want to transmit that information to and from their PCs.

Until recently, fax machines and personal computers had only a paper connection. Information produced with a personal computer had to be printed, taken to the fax machine, scanned and transmitted. At the other end, the information was received by a fax machine. If receivers wished to store the information in their own computers, it had to be re-entered through the keyboard. The system was slow, tedious and inconvenient, particularly for people who generated complex or lengthy documents.

ELECTRONIC MAIL VS. PC/FAX

Electronic mail—currently the major medium for computer communication—can be easy once you master it, but learning how to telecommunicate is not a simple task. The communications protocols in both the sending and receiving machines must match, and time delays built into the different telephone systems around the world follow no particular pattern. Therefore, an important transmission from Hong Kong might never get through, because the sending modem could hang up if the Hong Kong telephone system paused for a moment.

Setting up an effective telecommunications network often requires lengthy, expensive, individual consultation. Even a clear understanding of the details of telecommunication doesn't ensure that messages will be transmitted accurately if there's line noise.

Telecommunicating graphics information is even more complex. Graphics file formats often are incompatible, and not all communications software supports all graphics formats. Indeed, to send graphics information by electronic mail, software used by both sender and receiver not only should be compatible but identical. Unfortunately, that isn't usually the practice. As a consequence, even people who use electronic mail (E-mail) routinely for written documents generally send graphics information by overnight mail.

Because fax sends all information in a standard graphics format, it was inevitable that fax technology be integrated with computer technology not only to compete with E-mail but to allow instantaneous communication of computer-generated graphics. As an added benefit, PC/FAX conveniently links personal computer users with the expanding fax community.

With a burgeoning market, it's not surprising that more than 30 manufacturers now produce PC/FAX technology.

UNDERSTANDING PC/FAX

Although some companies offer PC/FAX in the form of an external box to be plugged into a PC's serial port, PC/FAX is most often an add-on board inserted into one of the computer's internal slots. With a board installed, a user can convert computer-generated documents into Huffman code—the code understood by fax machines—and send that information over regular telephone lines to any other fax machine in the world. In turn, it can receive information from any fax machine. Invariably, PC/FAX machines are Group 3-compatible. A small number also are Group 2-compatible.

An add-on board enables the computer only to send and receive fax messages. In many ways, a PC/FAX board is like a specialized modem, converting electronic data into a form that can be transmitted. However, to have your computer fully emulate fax machines, you must add a scanner to input externally generated documents. Your existing printer can be used to print incoming fax messages.

Fax Boards

A fax board generally takes either a whole or half slot, and a scanner may require a second slot. Because slots in PCs are at such a premium, some manufacturers offer fax boards integrated with both a high-speed modem and the scanner interface. At least one board also includes a small computer system interface (SCSI), so that other peripherals, such as a CD-ROM or laser disk drive, can be added. They may cost a bit more, but integrated PC/FAX boards are slot savers.

Optical Character Recognition (OCR) Versus Scanners

A scanner reads a document—text or graphics—dot by dot, makes a bit-mapped image and stores it as if it were a graphic. The information, therefore, can be manipulated only through graphics programs, provided fax software can convert your Huffman code into a compatible computer graphics format.

OCR readers start with a scanner and add either dedicated computer and software, or software only, to translate the text images from the document into text codes understood by the computer. The OCR software looks at each character in the document and compares it to known characters. When it finds a match, it assigns the appropriate code, called ASCII, to the character.

Consequently, text read into the computer via an OCR reader can be manipulated with such computer software as word processors. To date, the problem with inexpensive OCRs has been quality. Even a program that claims 99 percent accuracy could still miss a character on almost every line. Obviously that would require tedious editing for all documents read into the computer via OCR. Although expensive ($20,000-plus) OCR systems are more than 99 percent accurate; low-end systems are less reliable. Several companies sell add-on OCR software that can be used with many scanners, but this software is new and unproven.

Manipulating Fax Information

Most fax software packages include utilities that convert a fax image to other graphics file formats. After the conversion, the fax message can be changed by using a compatible PC paint program. Most boards minimally support PC Paintbrush and Dr. Halo.

However, the fax message is understood by the computer as a graphic—not as characters. For example, you can't receive a spreadsheet via fax and simply import the numbers to another spreadsheet for further analysis. As the new OCR add-on software discussed above matures, this limitation should also disappear.

Fax Board Software

As fax board technology has developed, the software that manages and operates the board has become an important consideration in evaluating different boards. Fax board software has two main functions: controlling the functions of the fax board and managing the interface between the fax board and other applications programs you may be running.

When you're choosing a board, consider the ease and convenience with which the fax board software controls the basic operation of the fax. A fax board should be able to do much more than simply convert an ASCII or graphics file into a fax file and transmit it to a sender within a couple of keystrokes; it should be able to perform advanced fax functions quickly and easily. For example, the software should let you develop an extensive directory of fax numbers and broadcast fax messages to a large number of recipients with minimum effort. Moreover, you should be able to construct groups to receive specific messages and to send different messages to different groups at different times.

The software should also let you check messages easily and view them on the screen, although many good fax boards take a long time to convert incoming fax files into screen-displayable form.

Another advanced function that PC/FAX board software should simplify is polling. By following four straightforward steps, you should be able to identify several files you want to have polled and set up the access procedure for that polling. The procedure should be simple.

Finally, the software accompanying a fax board should have standard utilities that can update the fax directory, review and edit the journal, reschedule events or cancel a transmission, and delete or rename a file.

Recent fax technology has moved well beyond the standard software included with a fax board. Several companies have added an interface between the fax software and popular applications programs, such as WordPerfect and Lotus 1-2-3, letting you move back and forth between them with a single keystroke. Furthermore, you can send the file you're working on as a fax without first having to convert it to ASCII or graphics format, or without exiting the applications program. The conversion process takes place automatically in the background. Some software even lets you create a standard cover sheet for files sent from an applications program. This makes a transmission sent via PC/FAX more closely resemble a standard fax.

Sending a fax while you're working in another application isn't entirely seamless, however. You may have to exit an application to slightly modify your setup, letting your computer know through which port to send a fax. Still, fax software is making it easier to use a fax board as an adjunct to other applications software. After all, the easier it is to use a fax board, the more likely you'll convert to PC/FAX.

Fax Board Prices

Depending on their features and specialized capabilities, fax boards range in price from $299 to about $1,495. If you already have a printer, be prepared to spend about $1,000 to $1,200 to add fax technology. On the other hand, when you turn your computer into a fax machine, you retain all the intelligence of the computer. Thus, for $1,000 or a bit more, you get all the features that you'd find in the most advanced fax technology, including a large amount of memory and complete programmability. The same functions in a stand-alone fax machine could cost as much as $3,000. And the price is dropping. At the time of this writing, a combination fax board and scanner was available for $1,000.

COMPATIBILITY ISSUES

PC/FAX is available for both IBM-compatible and Macintosh computers. Of course, once a board is installed, all Group 3 fax machines can communicate with each other, whether they're free-standing or reside in an IBM PC or a Macintosh. In other words, a fax in a personal computer can communicate with Group 3 faxes anywhere in the world.

Nevertheless, several compatibility issues remain with respect to hardware. First, the scanner must be compatible with the fax board itself. Many fax boards support only their manufacturers' scanners. Second, because fax information is graphic, it must be compatible with the computer's graphics board and monitor. Some people argue that CGA-level graphics are inadequate for heavy fax use and that EGA or VGA should be used.

To view received fax documents, you need a monitor with good display resolution. Resolution on computers is expressed as the number of horizontal and vertical dots that can be displayed on the screen. If you can't easily read an incoming fax message on screen, you'll first have to print each message on paper, which can be a slow task. Higher resolution graphics cards and displays make viewing faxes easier and faster by showing more of the document at one time on the screen. Hercules, EGA or VGA graphics are recommended for this use.

Even with good screens, you need more time to read a fax on screen than you would a standard fax transmission. Each page must be processed, creating pauses and delays. Often, each page must be resized to fit the screen as well.

Finally, the fax board must be compatible with the output device, generally the printer. Most manufacturers support Hewlett-Packard, Epson and PostScript protocols.

Ironically, you may run into some problems if you use a laser printer to output fax messages. Laser printers print at 300 by 300 dots per inch. Fax messages are sent at either 98 by 200 or 196 by 200 dots per inch. When you try to convert between resolutions that aren't even multiples, the image can be somewhat distorted. If you don't convert (and many boards aren't equipped with conversion software), a square sent in standard fax resolution will print as a rectangle.

Moreover, when you receive a message, you get only the type styles supported by the sending machine. This means that even if you've invested heavily in different fonts for your laser printer, they cannot be used to receive fax messages.

Other Hardware Considerations

Graphic images require large amounts of disk space. To convert a standard 8 1/2- by 11-inch image into 200 dots per inch requires nearly half a megabyte of storage. Even after the data are compressed, it can still take up 50K in storage space. Therefore, you need to reserve at least three megabytes of space on your hard disk for PC/FAX.

Storage space becomes even more valuable if you plan to save your fax messages on disk. Because some fax boards cannot store a compressed version of the message, you may need a huge hard disk to store just a few messages. Because even compressed messages use 50K per page, you won't be able to store many messages on disk.

In essence, this means that PC/FAX can be used efficiently only to capture and print messages. Most users routinely convert messages to hard copy to prevent their hard disks from becoming overloaded with old fax material.

RECEIVING MESSAGES VIA PC/FAX

Although they've been on the market a relatively short time, a third generation of PC/FAX boards has already been developed. In the first generation, PC/FAX ran in the foreground. In order to send or receive a fax, you couldn't do anything else with the computer. It had to be "dedicated" to fax usage.

In the second generation of PC/FAX, fax operations took place in the background but still called on the computer's central processing unit (CPU). That is, a user could continue to work on a specific application, but the computer's performance could be degraded as the CPU was called on to perform fax tasks. Also, second-generation boards sometimes interfered with RAM-resident programs.

The latest generation of fax boards come with their own CPU, usually an 80186 chip, and 256K RAM. As a result, they work entirely in the background and never call on the computer's CPU. Consequently, the process of sending and receiving faxes can be completely invisible unless the user is in the middle of an application that calls heavily on the hard disk itself. In such a situation, some minor delays could occur. When a fax is being sent, the user is signaled by a set of red LED lights attached to the side of the monitor or another convenient place, much the same way that lights signal modem or disk activity.

Although background operation eliminates the need to dedicate a personal computer to fax use, a user must still go into the fax program and check the activity log to ensure that a message has been sent successfully or to retrieve a recently received message.

PRINTING PC/FAX MESSAGES

Even though fax boards can send and receive messages in the background, printing a message often takes place in the foreground. That means when you transfer a message to paper, the computer can't be used for anything else.

This is a serious drawback. Inbound fax messages are in a graphic format, so printing them can be an infuriatingly long procedure, particularly if the sender is also transmitting photographs. A 12-page message with several photographs, for example, can take 30 minutes to print, tying up your computer the whole time.

Finally, although PC/FAX boards work in the background, your computer usually has to be turned on to receive a fax. Consequently, if you don't want to run your computer 24 hours a day, you won't be able to use PC/FAX as a round-the-clock receiving machine.

INSTALLING PC/FAX

If you've ever taken the hood off your personal computer, installation should take less than an hour. Most boards come with their own diagnostics. Some even have instructions detailing what to do if you don't understand the questions being asked in the course of the installation routine.

LEARNING TO USE PC/FAX

For some boards, sending fax is a two-step process—point to a file, point to a telephone number and the fax is on its way. Interestingly, advanced fax features such as polling, broadcasting and auto-dialing seem easier to master and more convenient to use on a personal computer than on a stand-alone fax machine. And they're much easier to learn than advanced modem communications techniques.

Differences in Fax Board Operations

Different fax boards convert ASCII and graphics formats into fax codes in different ways. Some boards convert the information while the fax message is being sent or received. Others complete the conversion off-line. Although the latter method takes longer to send a fax message, multiple transmissions of the same message can be completed more quickly using this method.

Still other boards take a middle approach. They begin the calling and conversion process at the same time but convert each page before they send it, filling an on-board buffer with the next page of the message. As a result, the first page of the transmission moves slowly, but the pace for the rest of the pages picks up.

Another major difference is the ability to preview what the fax message looks like in fax format.

Management Reporting Features

Most new-generation PC/FAX boards come with built-in communications journal and activity log reporting functions. Typically, the activity log and communications journal automatically record the file that was sent, the recipient (perhaps only their telephone number), the number of pages transmitted and the time it was sent. Unlike stand-alone fax machines, some fax boards omit the length of time the transmission required.

Other Software Features

PC/FAX boards tap the power of your computer

through software. In addition to management reporting functions and operations, the software generally includes a complete set of utilities to handle the fax information.

Different boards vary in capability. For example, if you want to broadcast a fax, the software on some boards lets you load in only ten telephone numbers at a time. Other boards have room for many more.

Hard Copy Reproduction

Interestingly, material generated on a personal computer and transmitted via PC/FAX often looks better at the receiving end than material sent by stand-alone fax. No matter how good a fax's scanner, there will always be some degradation in the image as it's scanned into the machine; and because PC/FAX generates the image internally, the scanning process is eliminated, making the image sharper.

For this reason, other fax machines used in tandem with PC/FAX can sometimes function as remote, high-quality printers. Suppose, for example, you have several offices with fax machines and computer equipment but your standard computer configuration has only nine-wire, dot-matrix printers or high-speed, draft-mode printers. If a message is generated on computer and sent via PC/FAX to a remote facsimile machine, it can be printed at nearly 200 by 200 dots per inch. In some applications, this may represent a huge leap in quality.

Recognizing that stand-alone fax machines can serve as remote printers for PC/FAX transmissions, some companies have devised boards that transmit AutoCAD and Hewlett-Packard Graphics Language (HPGL) drawings automatically via PC/FAX. An engineer who creates a drawing using AutoCAD and HPGL can send it to a remote location without ever printing it at the work site. The program automatically will cut the file into strips that can be reassembled to make a full-sized drawing printed at nearly 200 by 200 dots per inch at the receiving end. This reduces time substantially and provides significant convenience. Moreover, the drawing at the other end has a higher quality than if it had been scanned into a facsimile machine in the first place.

CAD/FAX, as it's called, works only when both machines are equipped with compatible boards. It's the first in a series of specialized applications that fax board manufacturers plan to develop and market over the next several years.

Unit Sizes with PC/FAX Transmission

Unfortunately, PC/FAX boards don't respect the page makeup protocols in computer software. Page breaks aren't read, and the information is sent in a fashion that corresponds to the internal logic of the fax board, rather than to the way the message looks on the computer screen. Nor do most fax boards add their own page numbers.

This limitation creates several problems. First, some fax boards don't allow the sender to preview a message before it's sent. For others, the routine to do this is clumsy and awkward. Consequently, with PC/FAX, the sender often doesn't know exactly what the receiver has received, a shortcoming that can be exacerbated by the error-correction schemes. For example, when some fax boards receive an error message after a page has been transmitted, the board will put the page aside, continue the transmission and only retransmit the faulty page at the end. If the pages aren't numbered, a great deal of confusion can result when the document is received.

Second, some fax boards automatically compress data generated in 132- and 200-column modes. Therefore, wide spreadsheets can be garbled.

FAX BOARD NETWORKING

Some fax boards are designed to work within networks, which can be very convenient for outgoing messages. Anyone on the network can create a document for fax transmission and send it to the board, which, in turn, will transmit the message to a fax machine anywhere in the world. Incoming messages, however, are less convenient. All incoming messages reside on one hard disk, and each recipient must check that disk periodically for messages. To receive and manage incoming fax messages adequately, the host computer must be equipped with a large hard disk and must be monitored regularly.

Some PC/FAX board vendors now offer products they claim will deliver a PC/FAX to a specific user on the network. Currently, this feature only works between boards made by the same manufacturer.

NEW DIRECTIONS

As microcomputer technology improves, fax boards continue to improve as well.

MCA/EISA

IBM's Microchannel Architecture (MCA) and the Enhanced Industry Standard Architecture (EISA), offered by IBM's competitors, were designed to boost the performance of add-on boards. Fax board manufacturers have begun to respond. For example, Gammalink now offers boards that can send up to eight faxes simultaneously when mounted in a computer with MCA capabilities.

Binary File Transfer

Another advantage of add-on fax boards is binary file transfer. Several boards can now transfer entire binary files of data contained in Lotus 1-2-3 spreadsheets or dBASE files as simple fax printouts. File transfer requires compatible boards at both ends, but since it transmits the data at 9,600 bits per second, it's faster than conventional telecommunications and more convenient than conventional fax.

Standard Communications Interface

As telecommunications, PC/FAX and other forms of file transfer find their niche, the need to link communications to specific application programs has grown. To allow software developers to incorporate a wide range of communications functions into their applications programs, Intel and Digital Communications Associates have developed Communicating Applications Specifications (CAS). The idea is to coordinate and simplify every aspect of PC communications, including PC/FAX.

CAS allows information to be sent to a fax board as if it were a printer. All communications functions can be controlled from within the application program. Intel offers CAS add-ins for Lotus 1-2-3 and WordPerfect 5.0 for people who buy its Connection CoProcessor, a combination fax board and modem. Several companies, including Microsoft, Ashton-Tate, Borland and Symantec have agreed to support CAS. Even without CAS, it's getting easier to use PC/FAX with application programs. For example, WordPerfect supports several fax boards with printer-drivers, which makes sending a document as a fax almost as easy as printing it.

THE BIG QUESTION: DO YOU NEED PC/FAX?

Not surprisingly, your need for PC/FAX depends on the way fax is used within your organization. PC/FAX is most efficient as a sending terminal if people who send fax messages generally create them on a personal computer.

For example, executives who write their own memos—or have them produced on a personal computer—may wish to have those memos faxed immediately to the receiver rather than wait for them to be printed and faxed by conventional means. The text image can be merged with a stored paint image of the company letterhead to produce a complete fax of the document.

Consultants, outside agencies and subcontractors, such as public relations and advertising agencies, who simply need customer approval for a job before they proceed, should explore PC/FAX if they create their roughs on a personal computer. Indeed, anyone who works primarily on a computer, generates heavy fax traffic, and wants to send his or her own faxes should explore PC/FAX.

PC/FAX is also for you if you can exploit its power and flexibility. Remember, PC/FAX eliminates most limitations of stand-alone fax with respect to the number of fax numbers that can be stored in a directory, unattended operation, delayed transmission and so on. Although Chapter 7 warns you of the dangers of junk fax, PC/FAX can generate and automatically send hundreds of faxes to get your message across, and it can customize those faxes to look just like any other fax traffic.

Remember, however, that a fax message is appropriate only if you assume that the recipient will be satisfied with a hard copy of the information. If you wish the receiver to have data that can be manipulated, it's usually more efficient to establish a direct telecommunications link.

At the same time, fax can be a substitute, although not a very efficient one, in situations where telecommunications links are impractical or are too difficult to establish. For example, it's extremely difficult to communicate overseas electronically with a modem. With PC/FAX, it's extremely easy to communicate overseas at any time.

PC/FAX also can be efficiently installed in a local area network. With one fax board, all members of the network can send their own fax messages whenever they want to, without relying on support staff.

On the other hand, PC/FAX doesn't serve equally well as a receiving machine, particularly when there's heavy fax traffic. Incoming messages on a PC not only require a lot of disk storage space, but may demand attention at an inconvenient time. In addition, the messages must be downloaded from your hard disk to your printer or into another storage medium. And unless the information is filed immediately, you risk losing it.

Finally, for your PC to serve efficiently as a receiving machine, it must run 24 hours a day. If you leave for a trip of any length, you must arrange for somebody to check your faxes regularly. Otherwise, you may not receive the information your communications partners are sending.

Clearly, within a business environment, PC/FAX has pros and cons. It's an effective sending machine, but requires a scanner in addition to the fax board to make it fully operational. As a sending machine, PC/FAX gives you additional control of your situation. As a receiving machine, it gives you less control because it demands too much attention at inappropriate times.

There's one situation, however, in which PC/FAX is invaluable—when you want to fax data which must be downloaded from a mainframe computer. With PC/FAX, the entire operation can be completely automated, requiring no user intervention. For example, in large retail chains, when an item is purchased, that information increasingly is sent from the cash register directly to a central computer that tracks inventory. Every night the central computer can download the performance and in-stock status of specific products to one or several personal computers that, in turn, can fax that information to the appropriate product manager or buyer. In distributing information that resides in large computers, PC/FAX offers tremendous efficiencies.

PC/FAX ALTERNATIVES

Sending faxes directly from a personal computer or communicating terminal is undeniably more convenient than printing the information before sending it. Since PC/FAX boards are limited in this area, both MCI Mail and Telenet have introduced international services that allow faxes to be sent via computer, without adding a board. The services work much like electronic mail (E-mail), but the messages can be received by Group 3 fax machines. All advanced features found in E-mail—such as the ability to send to large groups of people at the same time—are available with these services.

Since the costs are relatively low, the services can be a good alternative for people who want to send faxes from their computers but can't justify investing in a board. The only apparent drawback is that you can't receive fax messages from them; the information conversion process works only in one direction—from computer-readable information to fax information. This alternative is viable if you're familiar with dial-up services and you primarily need to send faxes, not receive them.

A CHECKLIST: IS PC/FAX FOR YOU?

1. Do you intend mainly to send information via fax or do you plan to receive fax as well?

2. If you don't plan to receive information via PC/FAX, how will you receive incoming fax? Bear in mind that once you begin to send fax, you're likely to receive fax in return.

3. Is the information you plan to send generated in personal computers? If it isn't, will you have to buy a scanner (which will add to the cost of the system)?

4. Can one computer station generate enough fax messages to justify the $1,000-plus investment? (It's as hard to share PC/FAX as it is to share a personal computer.)

5. Is the computer operator the right person to be a fax operator as well?

6. Is the information you want to send usable by the recipient in the form of hard copy? Would data be more useful?

7. Can you establish alternative electronics communications links with those whom you wish to
   communicate?

8. Do you have much experience with stand-alone fax?
   Can you identify specific advantages PC/FAX will
   provide?

The answers to these questions should lead you to the right
decision concerning PC/FAX. In many situations, PC/FAX
is the perfect solution to distribute fax power throughout the
office. Someday maybe all computers will have a fax board.

CHAPTER 7

THE CELLULAR/FAX CONNECTION

A mobile fax hooked into the cellular phone network seems the epitome of executive chic. The fax revolution has brought us the high-powered executive, riding in the back of a limousine, faxing instructions to subordinates around the world and receiving key business intelligence. The new power fantasy is a C.E.O. controlling a corporation from a yacht in the Caribbean, using fax to stay in charge.

Fax *can* be used in conjunction with cellular telephones, but mobile fax isn't nearly as easy, convenient or effective as standard fax transmission over telephone lines. In essence, cellular telephones are fancy two-way radios; to use fax or other data communications tools, a bridge must be built to link the radio and the telephone communications network, which isn't a trivial task.

Two ways of building those bridges have emerged. Several companies make couplers and jacks that link fax machines to cellular telephones; other products integrate fax machines and cellular interfaces into one package. And because fax machines need electrical power, devices are available that let fax machines tap into a car's cigarette lighter.

ACOUSTICAL COUPLERS

The first task in combining fax and cellular technologies is physically connecting the two worlds. Most fax machines hook into the telephone network via a standard RJ-11 port or plug, the kind of plug used with regular telephones, but not with cellular phones.

For cellular fax, the two types of physical connectors—jacks and acoustical couplers—were also used when modems were first introduced. A coupler fits over a telephone handset and transmits electronic tones. With it, a fax machine can be attached to virtually any phone in the world—cellular, PBX, pay phone, small key system—provided the machine is battery-operated or is near a convenient electrical plug.

Not restricted to cellular applications, the acoustical coupler enhances the machine's portability. With that in mind, a few fax manufacturers offer acoustical couplers with battery-operated, portable fax machines.

These fax machines are compact, no-frills units without document feeders or telephone handsets. They weigh about nine pounds and come with battery packs equipped with enough power to send about 25 pages of information.

These fax machines are adequate for sending or receiving short messages and can be convenient for travelers who can set them up in a hotel room. But connecting acoustical couplers to the cellular network has its problems. First, acoustical couplers transmit at only 2,400 bps (bits per second), one-fourth the speed of the standard 9,600 bps at which fax usually transmits. They also can allow extraneous noise to creep in. When a fax machine senses line noise, it automatically transmits more slowly. If the noise continues, the transmission might terminate or not go through at all.

Even under the best conditions, operating a fax machine acoustically coupled to the cellular network is an expensive proposition. Because cellular telephone users are billed for receiving as well as sending information, sending fax via this route is expensive, too. The transmission may not go through at all if there's a lot of noise. (Don't drive with the windows open when faxing!)

CELLULAR JACKS

A simpler way to link a fax machine to a cellular telephone is with a direct-connect jack—a box with a cable that plugs into a cellular telephone on one side and a standard RJ-11 port on the other. However, since cellular phones are really radios, not telephones, they don't have a dial tone; and fax machines need a tone to know when to answer a call or to dial out.

To overcome this problem, manufacturers offer two kinds of jacks—simple and smart. Simple jacks—which cost about half as much as smart jacks and are generally more compact—serve only as conduits to pass information from the fax machine to the cellular telephone. A person must be present to operate the fax machine. Even if it's an auto/receive machine hooked to a cellular telephone, the fax machine will sit idle without human intervention.

A smart jack allows a fax machine to work as if it were hooked up to the regular telephone network. By electronically providing an *off-hook tone*, a smart jack signals when the fax machine should answer or dial a call. A smart jack may also provide a dial tone, enabling users to attach telephone-answering machines and additional phone handsets. Finally, some smart jacks can produce standard telephone tones (DTMF), which are needed for placing credit card calls.

Which is better? Although smart jacks are more flexible and provide more options than simple jacks, particularly in remote field settings, they can be too expensive and too draining for a car battery to receive unattended fax transmissions via a cellular telephone. And because most people don't put their fax machine on auto/receive while they're driving (presumably most calls will be regular phone calls), they must remember to switch on the auto/receive function before they get out of the car.

Depending on the specific use, each jack has its advantages. With a jack, any small fax machine can be plugged into a cellular telephone. However, few manufacturers offer machines that weigh less than 20 pounds and have built-in cellular jacks.

Finally, jacks provide a faster rate of transmission than acoustical couplers.

Still, even with direct-connect jacks, most cellular-network fax transmissions travel at only 4,800 bps. And the fading and signal noise associated with all cellular communication is interpreted by a fax machine as line noise. Although the communications protocols built into fax machines protect data integrity and most data will get through, cellular fax transmissions aren't as fast or reliable as those sent via the standard telephone network.

POWER SUPPLY

The final component in mobile fax is the power supply. The battery packs that come with small portable fax machines are cumbersome and don't supply enough punch for demanding mobile applications. Some can handle only 15 pages of transmissions before they need to be recharged.

As a result, many suppliers offer power packs, called *inverters*. These devices convert a cigarette lighter's 12-volt DC electricity into the 120-volt AC current needed by fax machines. In addition to supplying power, some inverters signal and shut off when the car battery begins to run low. Others have built-in heat protection and surge protection.

MOVING ON

As the mobile office becomes more popular, the use of mobile and cellular fax will grow as well. The cellular telephone was the initial step in linking the traveling professional with his or her home office; cellular fax is the next.

Integrating fax and cellular technologies still has a long way to go. Fax transmissions via a cellular network are much slower, and the line noise associated with cellular continues to be a problem. Cellular and mobile fax are improving, but they have yet to match fax equipment used on a regular telephone network.

CHAPTER 8

INNOVATIVE FAX USE

With more than 2 million machines in place, fax use is widespread and should actually become commonplace in the next three to five years. Not only are businesses installing fax hardware, but telephone companies, seeing fax as a good opportunity to increase network usage, are offering such special promotions as free fax machines to businesses that switch to their long distance service.

Still, not everyone who uses fax realizes its full potential. In fact, the key question these days is not whether you use fax but how you use it. Fax should be more than just a communications tool; it should be a strategic weapon in your business plan. Used innovatively, fax machines can help you serve your customer better and perform more efficiently than your competition.

To take full advantage of fax, you should consider new fax capabilities; ways other companies are using fax that might apply to your business; and the problems in your business that might be solved by fax communication.

FAX ON THE ROAD

Perhaps the most intriguing new fax capacity is the commercial fax networks, which have begun to expand even beyond the public telephone network. The U.S. postal service has installed fax machines in more than 250 locations; and many business service centers and private mail services provide access to fax, too.

Fax machines are also being integrated directly into the transportation system. For example, next year Japan Air Lines plans to install fax machines in several of its jets. Pan American Airways is putting fax machines in its terminal lounges. A luxury train running between New York and Chicago is equipped with fax machines. Even the space shuttle Columbia has a fax machine on board!

The ability to send faxes from anywhere addresses some pressing business concerns. Ten or 20 years ago, most business travel was done by salespeople. Today, all levels of business professionals live out of suitcases at one time or another—and one of their most vexing problems is staying in touch both with the home office and with people they're traveling to see.

Let's say you're a business person on the road, and you have to reschedule your meetings. If you phone ahead to your clients, you often must leave a message with a secretary or co-worker. You have little chance to explain a schedule change or leave extensive new information.

Sending a fax helps resolve those problems. Even a short fax is better than a pink While-You-Were-Out slip.

Calling the home office can be even more troublesome, since you may have an extensive list of instructions for associates and subordinates, but probably don't want to spend a lot of time trying to reach them on the phone. And issuing instructions from a pay telephone at an airport between flights is an almost guaranteed way to garble some information.

The commercial fax network is beginning to change all this. As the network grows, you'll be able to send detailed messages to people who don't even have to be in their offices to receive them.

To take advantage of this benefit, you should carry fax numbers with you, to avoid having to call a company to get its number. If you have to call, you might as well not send a fax at all!

Also consider taking transmittal sheets with you. You'll probably want them to have a message area so you can send one-page transmissions. Having your own transmittal sheets maintains your company's professional image when you send a fax from a commercial fax center. Having prepared forms also cuts down on the steps involved in sending a fax.

The possibilities of hard-copy messaging are so promising that one company has devised a small machine that works via the building's electrical system. About the size of an adding machine—and using similar sized paper—these machines let you type short messages and send them electronically to other floors in the same building. The company claims that the unit saves people several hours a week in "telephone tag."

Hard-copy fax conversations may become as common as messages on large E-mail networks. Like conversations on-line, conversations via fax don't have to be conducted in real time, in other words both parties don't have to be present at the same time. Hard-copy conversations offer another advantage: an exact record of who said what to whom.

Of course, people sometimes may not want that exact record. In the 1989 gubernatorial campaign in Virginia, both the Democratic and Republican candidates conducted fax conversations with several newspapers and television stations, refuting the other's charges.

In addition to hard-copy messaging, the burgeoning fax network can give you a competitive advantage in other ways. For example, an internationally recognized consultant spends about two weeks of every month on the road. He often works on a portable computer armed with a fax board until 1:30 AM in his hotel room, recapping discussions and setting an agenda for the next day.

At 7:30 the next morning, he faxes a document to the client he is visiting that day. He then makes several copies when he arrives at their office. In essence, he uses the client's fax machine as a remote printer, so that he doesn't have to travel with a portable printer.

The use of hard-copy messaging and fax machines as remote printers help people operate more effectively on the road.

DISTRIBUTING INFORMATION

Many companies use fax machines as an alternative way to quickly distribute timely information to a select group of clients. A good example of this is the proliferation of faxed newsletters.

For example, twice a week the Swiss Bank of New York publishes a newsletter detailing international interest rates, which it sends via fax to 750 clients. Cambridge Energy Research also transmits a newsletter on worldwide fossil fuel reserves to hundreds of subscribers with PC/FAX. The trade magazine *Ad Week* has launched a fax newsletter as well.

Faxing a newsletter provides added impact with little added cost. The downside is that the publication's appearance depends, in large part, on the quality of the receiving machine; so it's best used only if the timeliness of the information truly warrants fax. If not, clients may resent having their fax machines tied up for long periods of time to receive a newsletter.

But directly distributing a newsletter by fax is by no means the most innovative use of the technology. The *Hartford Courant* and other newspapers have begun to fax a synopsis of the daily news to select subscribers, letting them preview the next morning's edition. Other companies use fax as a teaser—an advertisement for a larger packet of information being sent by mail or other communication channels. For example, the publisher of an on-line daily newsletter offers a weekly summary of that week's news via fax. The faxed copy serves two purposes: it helps sell the newsletter to people who don't have regular access to a computer; and it motivates people to subscribe to the daily version by giving them a peek at what they're missing by not going on-line. Interestingly, because the newsletter is on-line, the appearance of the faxed version isn't as important as it is in standard newsletter publishing.

Some newsletter publishers have used the interface between facsimile and computer technology to differentiate their product in other ways. For example, one daily on-line newsletter asks its readers to specify the information they want to read. The publisher then prepares a newsletter covering those topics and faxes it to the requesting subscriber. Fax subscribers not only receive the information as quickly as on-line readers do, they don't have to go on-line or wade through a lot of information that doesn't interest them.

Another company uses the opposite approach. Instead of faxing a condensed package of information to subscribers, it sends a larger one, but in a different form. Located in Washington, D.C., this father-son team collects as many press releases as possible in a given day and then faxes them, unedited, to their clients. The clients can cull the material for information they want—rather than relying on the little information that will make it to the daily newspaper.

Newsletters aren't the only type of information that can be effectively distributed by fax. It's no secret that many Americans move every year. As April 15 draws near, some need income tax forms from the state in which they used to live. Unfortunately, not having the proper form you need isn't a valid excuse for not paying your taxes! But now, with fax, anxious taxpayers can rely on a company that collects thousands of obscure forms from virtually every state and will fax accountants the forms they need. Because most people don't realize they're lacking an important form until the filing deadline is upon them, the ability to send the forms right away makes that enterprise particularly effective.

Each business mentioned above uses fax technology to distinguish it from the competition and improve its service. Via fax, you can send the same information you already do; customize information or send it in raw form; or send information that can't be sent cost effectively in any other way. The point is that fax can expand the way you use information.

FAX FOR FUN

Although much has been said about the need to limit fax usage to business purposes, fax is slowly emerging as a personal communications tool as well. In the "Cathy" comic strip, for example, single people fax their biographies to each other, hoping to arrange dates. Although it was meant to be funny, the idea isn't so far-fetched. In real life, one hi-tech Romeo faxed his picture and vita to more than 400 companies and received several marriage proposals. The offers, of course, came via fax.

Actually, fax has begun to replace short notes and memos of all kinds. That's not surprising, since more and more people have access to machines and fax is so easy and convenient that people often use it on impulse.

The use of fax is limited only by people's imaginations. Children with home fax machines have begun to fax homework and class notes to friends. One set of proud parents recently faxed a photocopy of their newborn baby's footprints as a birth announcement.

So many people are using fax for personal communications that the "Miss Manners" etiquette column has guidelines for when faxing is polite and when it isn't. Faxing a thank-you note for a gift sent by courier is acceptable; faxing notes for other gifts is a *faux pas*. The jury's still out on whether a faxed acknowledgement followed by a telephone call equals a written thank-you note.

FAX MISCELLANY

The common denominator among businesses using fax creatively is that they're actively seeking ways to expand and improve their services. Sending and receiving orders are routine fax applications; several mail order companies now accept orders via fax. Many companies have found that fewer mistakes are made with faxed orders and that for some reason, the average order is larger. This may motivate companies that don't generate a large volume of either written or telephoned orders to consider implementing fax.

Fax is also making it easier to get the right information to the right place at the right time. For example, some fire trucks are being equipped with cellular telephones linked to fax machines so they can receive the floor plan of a building as they rush to the scene of a blaze. The firefighters can develop their strategy while the sirens are still blaring. The advanced planning makes the firefighting more effective, and safer.

Fax has become a political tool as well. When political unrest erupted in China and Panama, fax was one form of communication that both governments found very difficult to control. When General Noriega closed down the major news media in Panama, Panamanians living in the United States faxed information into the country. In China, the opposite happened. When the Beijing government switched off the international broadcasts of the events in Tiananmen Square, Chinese students continued to fax information out.

MOVING ON

Many companies are still limited in their use of fax. Although it's important to ensure that fax is used responsibly, it's also important to look for new, innovative ways to use fax. For example, you don't need to add much to your own fax infrastructure to begin to use fax as a traveling communications system.

If you don't send newsletters or other regularly published

information to clients now, perhaps you should. Even a page or two every quarter, filled with chatty information designed to keep your name in your clients' minds, has been used effectively by many companies. A faxed newsletter is no more likely—and probably less likely—to be tossed aside without being skimmed than one received by mail. The information in a faxed publication can be more timely and the distribution much easier.

With new laws being passed in some states prohibiting junk mail, you'll have to get your clients' consent to fax a newsletter to them. But that will help build a closer relationship.

The first phase of the fax revolution is drawing to an end—people are buying the hardware. But the creative phase is only just beginning.

CHAPTER 9

A LOOK INTO THE FUTURE

Because fax technology will be an important element of our communication system for years to come, it's important to consider what the future might bring. Many "predictions" already exist, some of them through the combination of personal computers with fax. The real change will result when major fax manufacturers incorporate these enhancements into their main product lines.

RAPID BUT UNEVEN GROWTH

More than any other factor, growth will impact the future of fax. With fax machine sales rates projected to grow at 30 percent a year well into the 1990s, the sheer number of people with access to fax will shape its future most directly. As more people have fax, more people will need it. But at some point, perhaps within two or three years, growth will slacken off before the potential is fully realized. One buying group will be saturated, while another will have yet to enter the market. Nevertheless, even with more than two million fax machines in place in the United States, this initial stage of growth should be dramatic and impressive and help make fax more universal.

COMPATIBILITY

Group 3 machines will last well into the 1990s. Although you'll hear many promises about Group 4 (e.g., 5-second-per-page transmission speeds and totally error-free transmission), the use of Group 4 machines requires a new type of network. Because the telephone infrastructure won't change for many years, Group 3 fax is likely to serve you for a long time to come.

PRICING

Because so many companies are competing for a share of the market, fax buyers can look forward to continued pressure on manufacturers to provide more capabilities at the lowest price. Absolute prices may not drop far from current levels, but each price point should begin to offer more features for the money.

For example, today you can buy a stripped-down Group 3 fax for around $500; bottom-of-the-line fax computer boards cost about $200. At the same time, most buyers will probably opt for more feature-laden machines at a higher price.

The price categories outlined in Chapter 3 also will change. Because many small businesses want machines with a lot of features, manufacturers will try to offer as many features as possible below those price levels.

But price decreases and the changed price/performance ratios won't be sufficient to warrant deferring purchase. Because fax isn't like the typical consumer-electronics item, prices aren't likely to plunge. Indeed, fax probably will follow the pattern set by personal computers. Although PC clones now cost as little as $800, most people still spend more than $2,000 for a system.

ERROR CORRECTION

In the short term, one major enhancement to fax machines will be the addition of Error Correction Mode (ECM). The CCITT has established a Group 3 optional ECM standard that allows different brands of machines to communicate with each other and detect errors during transmission. As each block of text is sent, the receiving machine is able to determine whether it has correctly received all the information. If it detects an error, it automatically requests the sending machine to resend the block, resulting in an error-free message. New machines with ECM can be found in all but the lowest price range, but both sending and receiving machines must support it.

INCREASED PAPER CAPACITY, LASER PRINTING AND BOND PAPER

The quality of paper and printing are weak areas in fax technology today, and many improvements are likely. Thermal printing is inexpensive, rugged and quick, and it lends itself to fax. But it doesn't have the look, feel or durability of regular bond paper.

As fax machine volume increases and as copier equipment continues to drop in price, laser printing technology is likely to replace thermal printing. The new laser-based fax machines come with lower price tags. While older machines listed for $6,000, the new models cost less than $5,000. Within the next two years, discounts should bring that price down to $3,000.

Inkjet technology (such as that used in the Hewlett-Packard DeskJet printer, which sells for less than $800) is also a viable fax printing technology. It uses plain paper to print, and its speed, though slower, is adequate for fax use.

Large companies may begin to use fax units with multiple bins, so that internal memos can be printed on one type of paper or form and documents from outside the company can be printed on another.

INTEGRATION WITH OTHER OFFICE EQUIPMENT

When fax machines begin using laser printer technology, multipurpose machines will appear. Both computer laser printers and fax machines work efficiently in a group or departmental setting. A future combination fax/laser printer might have a port into which a document scanner can be plugged. The machine might be used in a local network to print computer documents or used as a sending and receiving fax machine.

Adding full-function copier capability is also a natural step. With document feeders and a scanner, your machine can be used for both fax and high-quality copying.

The internal memory used by fax machines also will be useful for copier functions. Original documents need to be scanned only once; then the machine can print as many copies as needed from stored memory, resulting in less wear and tear on originals. Multiple output bins will be even more important as new multipurpose machines appear.

Although fax machines have telephone handsets, the trend isn't toward two-line units, which would allow the handset to be the primary telephone. Fax will continue to be a convenient and accessible technology, but not so accessible as to jeopardize security or to do double duty as a substitute telephone system.

INCREASED MEMORY

More memory will increase the flexibility of fax usage by allowing easier implementation of such advanced features as delayed transmission and broadcasting. Memory will be calculated in pages of information, and 100-page memories won't be uncommon.

Memory also will be used to solve fax security problems. When a document is received, only a cover sheet will be printed for the recipient. After the recipient enters a security code on the machine, the document will be printed from the fax machine's internal memory. This feature is available today on high-end machines and someday should be available on midrange models, too.

NEW COMPUTER-BASED MACHINES

Because computers will remain the center of office automation, PC/FAX will continue to proliferate. In addition to PCs with fax capability, a PC-based fax machine that performs only fax functions may appear, much like the Magnavox Video Writer, a PC-based dedicated word processor.

The trend in PC/FAX will be toward multifunction and multiport communication boards. These new boards will include fax features as well as computer modems, document-scanner ports and even ports for video images which will form the basis of communication boards. Optical Character Recognition (OCR) will evolve into an integral part of PC/FAX. Once a fax is received, it will be converted to text data for storage or further processing. The ability to convert from image to data is a key element for the viability of PC/FAX. In the future, you'll be able to see the people you talk to on the screen and, at the same time, send faxes or other data on the same machine. This technology is now being developed by some PC/FAX board manufacturers.

INTEGRATION WITH OTHER FORMS OF COMMUNICATION

As computer-to-computer electronic mail becomes increasingly important in large corporations, links to technology such as fax will become more common. Both AT&T Mail and MCI currently provide computer electronic-mail services to large companies.

For example, a buyer in a large company can fax requests for proposals to three vendors and send an E-mail copy of the same request to his or her boss. The text can be entered only once, and the destination can be a computer or fax machine. Together, fax and E-mail will replace telex.

"FOURTH GENERATION" FAX

ISDN, which stands for Integrated Services Digital Network, will be adopted by the telephone system during the 1990s. But it has a long way to go before most of us will be affected by it.

Today's telephone network sends analog signals along phone lines. The new ISDN network will be designed for digital information, such as that used by computers and fax machines. In many ways, the difference between the two technologies is like that between phonograph records and compact disks (CDs). Records are analog, CDs are digital; and although CDs produce unmatched sound quality, they require new equipment.

Given the current size of the telephone network, the new equipment for ISDN will take many years to install. However, once ISDN is here, new Group 4 fax machines will be able to make use of its high-speed digital lines to speed error-free faxes in less than five seconds per page.

PORTABLE FAX

As more and more businesses rely on fax communication, portable fax machines will become popular with traveling executives. With a portable fax machine, travelers will be able to receive and review reports and memos in their hotel rooms. Field sales representatives will use portable fax machines together with cellular phones to send and receive information from their cars.

"Portable fax" can mean anything from small lightweight machines requiring AC power to even lighter battery-operated units. Like laptop computers, portable fax will proliferate as people learn new ways to use it.

THE FUTURE OF FAX

Through professional education, thousands of new fax users will understand and then utilize fax in novel settings and situations. Companies will increase their fax applications, but the technology won't be restricted to the office environment. For example, as purchasing managers in various industries and professions rely on fax to obtain price quotes and to place orders, more and more suppliers will add fax capability to better serve their customers.

Contractors may use fax to send documents to the building commission and get back comments or building permits without waiting for hours at city hall. Emergency prescriptions may be faxed to pharmacies, eliminating telephone calls and providing verification.

Since it could be abused in a number of ways, fax isn't likely to become popular for home use. Home fax machines would be overrun with contest offers and junk mail solicitations.

Because fax is less expensive than mail in a local telephone area, many companies would bill through fax if it were widespread. How would you feel if you paid for the fax paper each time you received a bill and didn't get a reply envelope for your check? The requirement for a dedicated phone line will also greatly limit home fax use.

At the office, although different departments will demand their own machines, faxes will not be found on every desk as telephones are today. PC users will use PC/FAX more and more if they're creating and sending large amounts of information. And personal faxes will be used by high-level executives (or their staff) who work with sensitive data and need to keep the information confidential.

As long as paper remains a convenient medium for sending and storing information, fax will continue to grow as an important, even necessary, communications technology. But it will grow in an evolutionary fashion. The essential process—the ability to scan, send and receive information on hard copy via telephone lines quickly and inexpensively—is already in place.

Although fax technology will improve, there's no reason to wait for the next generation of equipment if you have relevant fax applications now. In most cases, companies will add to their fax network incrementally, with international standards ensuring that current fax technology won't become obsolete. Although you can expect fax technology to be improved five years from now, the current state of technology is good enough to have a dramatic and beneficial impact on your communications infrastructure.

TO LEASE OR TO BUY

As with other types of office equipment, many companies opt to lease fax machines rather than buy them. Most major suppliers provide leases for their larger machines, and office equipment dealers generally offer leases through third-party leasing companies.

There are two major types of leases—a financial lease and an operating lease. A financial lease resembles traditional borrowing. The lessee makes a series of payments which at least equal the full price of the machine. The lease can't be cancelled. At the end of the lease, the lessee owns the machine.

Financial leases are popular with small businesses for several reasons. Because lessors are, in essence, lending money against a specific piece of equipment (the fax machine), and because the lessors own the machine until the lease ends, lessors are better protected against nonpayment than general lending institutions. Consequently, it may be easier for a small business to lease equipment than to borrow money to buy the machine.

A business usually can get a lease for 100 percent of the selling price of the equipment. However, the rates on a financial lease may be higher than the rates offered by other lending sources.

The second type of lease, an operating lease, most resembles equipment rental. An operating lease protects a company against the obsolescence of equipment, because operating lease payments typically don't cover the full price of the machine. At the end of the lease period, however, the lessor still owns the equipment and can lease or sell it to another party, while the lessee can choose a more advanced model. While most operating leases can be cancelled, the cancellation terms can be so onerous that buying the machine outright in some situations is more practical.

Leasing also has tax implications. For example, on leased machines, companies can take a research and development tax credit which otherwise isn't allowed. Moreover, companies subject to the alternative minimum tax can accrue certain tax advantages through leasing. Because every company's situation is different, consult with your accountant to determine your course of action.

In evaluating leasing options, remember that a basic Group 3 fax retails for less than $1,000, and a top-of-the-line unit with all the bells and whistles may cost $5,000. As a result, it can be more efficient just to borrow the money to buy a machine, or a few machines, outright.

COMMERCIAL FAX CENTERS

Despite the explosive growth of fax, many companies still don't have fax machines. Indeed, you may have decided that you don't wish to invest in fax technology right now. Your volume of fax traffic may be too low, or you may want to wait for the next generation of machines.

To meet the needs of those who only occasionally use fax, many business service centers, mailbox kiosks and quick copy centers have begun to offer fax facilities to the public. Though expensive, the services are easy to use. To send, you simply bring your document to the center and an operator does the rest. To receive, you simply tell the sending party the telephone number of the fax center you plan to use. Be sure the sender includes your name and telephone number on the transmittal sheet.

As demand increases, the price for fax services will fall dramatically. The services generally use one of two billing methods. The first is a per-page charge that includes the telephone charges that the service incurs. Some typical rates are as follows:

Incoming faxes: $1 per page.

Outgoing faxes:

> a) within the U.S. and Canada—$2 per page for the first three pages, $3 for each subsequent page (the transmittal sheet counts as a page, and "pages" means standard 8.5- x 11-inch letter-size paper; 8.5- x 14-inch, legal-size pages cost 50 percent more).

> b) international—$10 for the first page, $7 for the second and $5 for each additional page.

The second method of billing used by fax services is a per-page rate for the service and a separate bill for the associated telephone charges. With this method, typical rates for sending or receiving are $2 for the first page and $1 for each additional page, plus phone charges.

For someone whose fax use is limited and infrequent, a fax service provides an alternative to owning a fax machine. It also allows you to fax materials to people who don't own a machine. Clearly, for documents of five pages or less, even fax services represent a competitive alternative to overnight mail. On the other hand, if you routinely use a fax service, you may want to buy your own fax equipment.

SELECTED FAX RESOURCES

Vendors/Manufacturers

Abaton Technology
4831 Milmont Dr.
Fremont, CA 94538
415/683-2226

AT&T
Parsippany, NJ 07054

Canon USA, Inc.
One Canon Plaza
Lake Success, NY 11042
516/488-6700

Cypress Research
766 San Aleso Ave.
Sunnyvale, CA 94086
408/745-7150

Datacopy Corp.
535 Oakmead Pkwy.
Sunnyvale, CA 94086
408/245-7900

Dataquest, Inc.
1290 Ridder Park Dr.
San Jose, CA 95131
408/437-8000

Gammalink
2452 Embarcadero Way
Palo Alto, CA 94303
415/856-7421

Konica Business Machines
500 Day Hill Rd.
Windsor, CT 06095
203/683-2222

Lantor, Inc.
13650 Gramercy Pl.
Gardena, CA 90249
213/324-7070

Murata Business Systems
5560 Tennyson Pkwy.
Plano, TX 75024
214/403-3300

NEC of America
8 Old Sod Farm Rd.
Melville, NY 11747
516/753-7000

Panasonic
Data Communications Products
Two Panasonic Way
Secaucus, NJ 07094
201/348-9090

Pitney Bowes Corp.
1 Elmcroft Dr.
Stamford, CT 06926-0700
800/672-6937

Q/COR
1 Meca Way
Norcross, GA 30093
404/923-6666

Radio Shack
A Division of Tandy Corporation
1700 One Tandy Center
Fort Worth, TX 76102
817/390-3011

Ricoh Corp.
5 Dedrick Pl.
West Caldwell, NJ 07006
201/882-2000
800/727-4264

Sanyo Business Systems Corp.
51 Joseph St.
Moonachie, NJ 07074
201/440-9300

Savin Corp.
9 West Broad St.
Stamford, CT 06904-2270
203/967-5000

Sharp Electronics Corp.
Sharp Plaza
Mahwah, NJ 07430
201/529-8200

Tech Data Corp.
5350 Tech Data Dr.
Clearwater, FL 34620
813/539-7429

TEO Technologies
#60 Mural St. Unit #7
Richmond Hill, ONT L4B 3H6
Canada
416/889-0110

Toshiba America
Irvine, CA 92713-9724
714/583-3000

Xerox Corp.
Parts Marketing Center
Xerox Building 214-07S
P.O. Box 1020
Webster, NY 14580
800/828-5881

Research

BIS CAP
P.O. Box 68
Newtonville, MA 02160
617/893-9130

Camarro Research
P.O. Box 691
Fairfield, CT 06430
203/255-4100

Cores Corp.
767 Third Ave., 34th Floor
New York, NY 10017
212/888-0188

Datapro International
600 Delran Pkwy.
Delran, NJ 08075
609/764-0100

Dataquest
1290 Ridder Park Dr.
San Jose, CA 95131
408/437-8000

Accessories and Supplies

Paper Manufacturers Co.
24 Triangle Park Dr.
Cincinnati, OH 45246
800/327-4359
In Ohio 800/423-8512

Curtis Manufacturing Co.
30 Fitzgerald Dr.
Jaffrey, NH 03452
603/532-4123

Demco, Inc.
4810 Forest Run Rd.
Box 7488
Madison, WI 53707
608/241-1201
800/356-8394
Fax 608/241-1799

Electronic Speech Systems, Inc.
1900 Powell St., Ste. 205
Emeryville, CA 94608
415/547-2755

Lanier Worldwide, Inc.
2300 Parklane Dr., N.E.
Atlanta, GA 30345
404/496-9500

ONEAC Corp.
27944 North Bradley Rd.
Libertyville, IL 60048
800/533-7166

Publications

First FAXts
Dataquest
1290 Ridder Park Dr.
San Jose, CA 95131
408/437-8000

Geyer's Office Dealer
51 Madison Ave.
New York, NY 10010
212/689-4411

Modern Office Technology
1100 Superior Ave.
Cleveland, OH 44114
216/696-7000

Office Products Evaluation
600 Delran Pkwy.
Delran, NJ 08075
800/328-2776

Office Products Dealer
191 S. Gary Ave.
Carol Stream, IL 60188
312/665-1000

The Authorized OA Dealer Report
580 Commerce Dr.
Fairfield, CT 06430
203/336-4566

The Office
1600 Summer St.
Stamford, CT 06905
203/327-9670

What to Buy for Business
350 Theodore Fremd Ave.
Rye, NY 10580
914/921-0085

Directories

Dial-A-Fax Directory, third ed.
Dial-A-Fax Directories Corp.
P.O. Box 668
Jenkintown, PA 19046
800/346-3329

**The Directory of Telefacsimile Sites in Libraries
in the United States and Canada**
CBR Consulting Services
Box 248
Buchanan Dam, TX 78609-0248
512/793-6118

Facsimile Users' Directory
Monitor Publishing Company
104 Fifth Ave.
2nd Floor
New York, NY 10011
212/627-4140

Fax-Net International
Box 825
Flushing, NY 11354
718/463-2100

Feature Reports
Datapro International
600 Delran Pkwy.
Delran, NJ 08075
609/764-0100

Spec Check Facsimile Guide
Dataquest
1290 Ridder Park Dr.
San Jose, CA 95131
408/437-8000

The NOMDA Spokesman
12411 Wornall Rd.
Kansas City, MO 64145
816/941-3100

Associations

American Facsimile Association
1701 Arch St.
Philadelphia, PA 19103
215/568-8336

International Fax Association
4023 Lakeview Dr.
Lake Havasu City, AZ 86403
602/453-3850

National Office Machine Dealers Association
(NOMDA)
12411 Wornall Rd.
Kansas City, MO 64145
816/941-3100

Index

A

Acoustical couplers 136
Alternate-number dialing 31
ASCII 118, 120, 125
Automatic
    fallback 36
    paper-cutting 34
    redial 30
    voice/data switch 24, 32, 63

B

Backup 100
Batch index 37
Baud rate 19, 25, 36
Broadcast transmission 41, 42, 62, 125
Built-in dialing 27

C

Call notification 32
Call waiting 87
CCITT 17, 151

Cellular 135–139
  jacks 137
  telephone 135
Coding 117, 120
  modified Huffman 18–19
  read 19
Commercial fax centers 142, 159–160
Communications journal 47
Communications options 3
  computer networks 7
  fax 8
  messenger service 5
  telegram 6
  telephone 4
  telex 6
  US mail 3
Compatibility 16, 54, 121, 150
Computer networks 7
Confidential fax 64, 82, 83, 102
Contrast control 45
Copier 58, 152
Copier option 35
Cost management 66, 112
Costs 53, 64, 67, 120, 150, 159–160
Cover sheet 97

D

Data
  compression 25
  decompression 25
Delayed transmission 32, 34, 40, 57, 61, 110
Delivery 104
Dial cards 30
Discount and mail-order 72, 89
Distribution 103
Document
  conversion 125
  feeder 34, 40

feeder capacity   42, 57, 61
reduction   45
size   23, 45
Document size   57
Dot matrix printer   22, 47, 126

E

Electronic mail   116
Error correction   36, 44, 127, 151

F

Fax
abuse   107
advanced features   26
advances   128–129
advantages   8
announcement   95
basic features   16
boards   117, 119–123, 128–129
commercial centers   141–148, 159–160
communications activity   66
communications options   3
compatibility   See Compatibility.
costs   10
directory   29, 105, 110
disadvantages   8
for fun   146
growth   149
installation   87
junk mail   104
leasing   157–158
machine location   82
machine size   26
machine weight   26
management features   47
manager   90, 110
messaging   143

miscellany 147
misconceptions 2
number 29, 86, 95, 105, 110
operation 24
paper 22
portable See Portable fax.
See Store and forward.
testing 91
See Transmission.

G

Graphics 116
Group 1 17–18, 55
Group 2 17–18, 68, 117
Group 3 17–22, 46, 53–59, 67–68, 117, 121, 150
    resolution 19–22
Group 4 17–21, 70, 150

H

Halftone transmission 46, 59
Handshaking 60

I

Image enhancement 44
Inappropriate faxes 100
Information management 94
Inkjet printing 151
Installation 87, 124
Internal memory 57, 61
ISDN 17, 19, 70, 154

J

Junk fax 104

L

Laser printer   20–21, 66, 111, 122, 151
LCD display   29
Lease versus buy   157
Legal issues   108

M

Machine location   82
Maintenance   22, 111
Management features   47, 125
Manual receive   27
Manufacturers' sales reps   71, 88
Memory
    features   41
    See Internal memory.
    transmission   43
Messenger service   5
Modem   25, 36, 68, 117

N

Networking   127
Newsletters   144

O

OCR   118, 153
Office equipment stores   72, 88
One-Touch dialing   28, 61
Operator   24, 102, 109
Optical mark reader (OMR) sheets   41
Overnight mail   66

P

Page numbering 37, 127
Paper 55, 151
  bond 22–23
  cutting 22, 34
  roll 22, 55
  sheet 23
  size 22–23
  thermal 22, 59
PC/FAX 115–133, 153–156
  alternatives 131
  costs 118
  disk space 122
  evaluating 129
  screen graphics 121
Personnel management 109
Polling 38, 57, 63, 119, 125
  free 39
  secure 39
Portable fax 26, 135, 141, 155
Power outage 90
Price groups 53, 67
Printer protocols 121–122
Production features 44
Programming 89
Protecting faxes 103

R

Report
  communications activity 47–48
  reception 56
  transmission 47
Resolution 19–22, 25, 59
  regular 20
Routing 102
RS 232 port 46, 66

S

Scanners   58, 118
Scanning width   58
Security   39, 64, 102
Sheet feeder   34, 40, 57
Software   118–119, 125
Speed dialing   29, 61
Store and forward   41, 43, 62
Stored-number memory   61
Storing messages   102
Superfast transmission   60
Surge protector   90

T

Telegram   6
Telephone   24
   cellular   135
   communications options   4
   dialer   27
   handset   27, 58
   line   27, 45, 46, 58, 85, 86
   local company   62, 81
   long-distance service   28, 85
Telephone line   84
Telex   6, 82, 108
Testing   91
Timer transmission   40
Training   88, 91, 111
Transmission
   audit   47
   See Broadcast transmission
   See Delayed transmission
   errors   36
   features   27–44
   modem   25
   report   47

speed   17–19, 25, 54, 60
 time   24, 47, 54, 60
Transmission
 errors   47
Transmit-terminal identification   36, 56
Transmittal sheet   97, 143
TWX   66

U

Unattended operation   24, 41, 56
US mail   3

V

Voice/data switch
  See Automatic voice/data switch

W

WATS   86
White-space skipping   25, 60

Enhance your productivity with more books from Ventana Press

TO ORDER ADDITIONAL COPIES OF
THE BOOK OF FAX

Please send me _____ additional copies of *The Book of Fax* at $9.95 per book. Add $2.40 per book for normal UPS shipping ($1 per book, thereafter); $5 for UPS "two-day" air. North Carolina residents add 5% sales tax. Immediate shipment guaranteed.

Note: 15% discount for purchases of 5-9 books. 20% discount for purchases of 10 or more books. Resellers please call for wholesale discount information.

Name _____ Co. _____

Address (no P.O. Box) _____

City _____ State _____ Zip _____

Daytime telephone _____

_____ Payment enclosed (check or money order; no cash please)

_____ Charge my VISA/MC Acc't # _____

Exp. Date _____ Interbank # _____

Signature _____

Ventana Press ■ P.O. Box 2468 ■ Chapel Hill, NC 27515 ■ 919/942-0220
FAX 919/942-1140 (Please don't duplicate your fax orders by mail.)

Please send me _____ additional copies of *The Book of FAx* at $9.95 per book. Add $2.40 per book for normal UPS shipping ($1 per book, thereafter); $5 for UPS "two-day" air. North Carolina residents add 5% sales tax. Immediate shipment guaranteed.

Note: 15% discount for purchases of 5-9 books. 20% discount for purchases of 10 or more books. Resellers please call for wholesale discount information.

Name _____ Co. _____

Address (no P.O. Box) _____

City _____ State _____ Zip _____

Daytime telephone _____

_____ Payment enclosed (check or money order; no cash please)

_____ Charge my VISA/MC Acc't # _____

Exp. Date _____ Interbank # _____

Signature _____

Ventana Press ■ P.O. Box 2468 ■ Chapel Hill, NC 27515 ■ 919/942-0220
FAX 919/942-1140 (Please don't duplicate your fax orders by mail.)

NO POSTAGE
NECESSARY
IF MAILED
IN THE
UNITED STATES

BUSINESS REPLY MAIL
FIRST CLASS MAIL   PERMIT #495   CHAPEL HILL, NC

POSTAGE WILL BE PAID BY ADDRESSEE

Ventana Press

P.O. Box 2468

Chapel Hill, NC 27515

NO POSTAGE
NECESSARY
IF MAILED
IN THE
UNITED STATES

BUSINESS REPLY MAIL
FIRST CLASS MAIL   PERMIT #495   CHAPEL HILL, NC

POSTAGE WILL BE PAID BY ADDRESSEE

Ventana Press

P.O. Box 2468

Chapel Hill, NC 27515

TO ORDER ADDITIONAL COPIES OF
THE BOOK OF FAX

Please send me _____ additional copies of *The Book of Fax* at $9.95 per book. Add $2.40 per book for normal UPS shipping ($1 per book, thereafter); $5 for UPS "two-day" air. North Carolina residents add 5% sales tax. Immediate shipment guaranteed.

Note: 15% discount for purchases of 5-9 books. 20% discount for purchases of 10 or more books. Resellers please call for wholesale discount information.

Name _____ Co. _____

Address (no P.O. Box)_____

City_____ State _____ Zip_____

Daytime telephone_____

_____ Payment enclosed (check or money order; no cash please)

_____ Charge my VISA/MC Acc't # _____

Exp. Date _____ Interbank # _____

Signature _____

**Ventana Press ■ P.O. Box 2468 ■ Chapel Hill, NC 27515 ■ 919/942-0220
FAX 919/942-1140 (Please don't duplicate your fax orders by mail.)**

Please send me _____ additional copies of *The Book of FAx* at $9.95 per book. Add $2.40 per book for normal UPS shipping ($1 per book, thereafter); $5 for UPS "two-day" air. North Carolina residents add 5% sales tax. Immediate shipment guaranteed.

Note: 15% discount for purchases of 5-9 books. 20% discount for purchases of 10 or more books. Resellers please call for wholesale discount information.

Name _____ Co. _____

Address (no P.O. Box)_____

City_____ State _____ Zip_____

Daytime telephone_____

_____ Payment enclosed (check or money order; no cash please)

_____ Charge my VISA/MC Acc't # _____

Exp. Date _____ Interbank # _____

Signature _____

**Ventana Press ■ P.O. Box 2468 ■ Chapel Hill, NC 27515 ■ 919/942-0220
FAX 919/942-1140 (Please don't duplicate your fax orders by mail.)**

BUSINESS REPLY MAIL
FIRST CLASS MAIL   PERMIT #495   CHAPEL HILL, NC

POSTAGE WILL BE PAID BY ADDRESSEE

Ventana Press

P.O. Box 2468

Chapel Hill, NC 27515

NO POSTAGE
NECESSARY
IF MAILED
IN THE
UNITED STATES

BUSINESS REPLY MAIL
FIRST CLASS MAIL   PERMIT #495   CHAPEL HILL, NC

POSTAGE WILL BE PAID BY ADDRESSEE

Ventana Press

P.O. Box 2468

Chapel Hill, NC 27515

MORE ABOUT VENTANA PRESS BOOKS . . .

If you would like to be added to our mailing list, please complete the card below and indicate your areas of interest. We will keep you up-to-date on new books as they're published.

_____Yes! I'd like to receive more information about Ventana Press books. Please add me to your mailing list.

Name _____

Company _____

Street address (no P.O. box) _____

City _____ State _____ Zip _____

Please check areas of interest below:

_____ AutoCAD   _____ Newsletter publishing

_____ Desktop publishing   _____ Networking

_____ Desktop design   _____ Facsimile

_____ Presentation graphics   _____ Business software

Please return the postage-paid card to Ventana Press, P.O. Box 2468, Chapel Hill, NC 27515, 919/942-0220, FAX 919/942-1140. (Please don't duplicate your fax requests by mail.)